Intelligent Algorithms

Intelligent Algorithms
Theory and Practice

HAN HUANG

School of Software Engineering, South China University
of Technology, Beijing, P.R. China
Key Laboratory of Big Data and Intelligent Robot (SCUT),
MOE of China, Guangzhou, P.R. China
Guangdong Engineering Center for Large Model and
GenAI Technology, Guangzhou, P.R. China

ZHIFENG HAO

College of Science, Shantou University, Guangdong,
P.R. China

ELSEVIER

Elsevier
Radarweg 29, PO Box 211, 1000 AE Amsterdam, Netherlands
125 London Wall, London EC2Y 5AS, United Kingdom
50 Hampshire Street, 5th Floor, Cambridge, MA 02139, United States

MATLAB® is a trademark of The MathWorks, Inc. and is used with permission. The MathWorks does not warrant the accuracy of the text or exercises in this book. This book's use or discussion of MATLAB® software or related products does not constitute endorsement or sponsorship by The MathWorks of a particular pedagogical approach or particular use of the MATLAB® software.

Notices
Knowledge and best practice in this field are constantly changing. As new research and experience broaden our understanding, changes in research methods, professional practices, or medical treatment may become necessary.

Practitioners and researchers must always rely on their own experience and knowledge in evaluating and using any information, methods, compounds, or experiments described herein. In using such information or methods they should be mindful of their own safety and the safety of others, including parties for whom they have a professional responsibility.

To the fullest extent of the law, neither the Publisher nor the authors, contributors, or editors, assume any liability for any injury and/or damage to persons or property as a matter of products liability, negligence or otherwise, or from any use or operation of any methods, products, instructions, or ideas contained in the material herein.

ISBN: 978-0-443-21758-6

For Information on all Elsevier publications
visit our website at https://www.elsevier.com/books-and-journals

Publisher: Matthew Deans
Acquisitions Editor: Glyn Jones
Editorial Project Manager: Naomi Robertson
Production Project Manager: Selvaraj Raviraj
Cover Designer: Christian Bilbow

Typeset by MPS Limited, Chennai, India

Working together
to grow libraries in
developing countries

www.elsevier.com • www.bookaid.org

Contents

About the authors

Han Huang

Dr. Huang is a Professor and Doctoral Supervisor of the School of Software Engineering at South China University of Technology. He is currently serving as an Associate Editor of *IEEE Transactions on Evolutionary Computation* (IF: 14.3), *Complex & Intelligent Systems* (IF: 5.8), and *IEEE Transactions on Emerging Topics in Computational Intelligence* (IF: 5.3) and Director of Teaching Steering Committee for Software Engineering of Undergraduate Colleges and Universities in Guangdong Province. Prof. Huang has made great contributions to the scholarship on the theories and application of intelligent optimization algorithms. For example, he has proposed a time complexity analysis method of real-world evolutionary algorithms, algorithms for efficient and accurate image matting, a method for automated test case generation based on path coverage, and so on. Prof. Huang has hosted more than 20 national and provincial projects. He has published two books, *Theory and Practice of Intelligent algorithm* and *Theory* and *Methods and Tools for Time Complexity Analysis of Evolutionary Algorithm*. He has also published more than 80 papers in *IEEE TCYB, IEEE TETC, IEEE TSE, IEEE TEVC, IEEE TIP, IEEE TFS*, and *Science China*, including ESI highly cited papers. As the first inventor, Prof. Huang has 48 invention patents granted in China and 7 invention patents granted in the United States. He won China Patent Excellence Award and developed an association standard entitled "Standard for glass-box testing without source code" as the first completer. Additionally, Prof. Huang pays attention to social services. Over the past 5 years, he has given more than 50 public lectures on science and technology for government offices, primary and secondary schools, CCF, YOCSEF, media, and so on. He has been in charge of the development and release of public software systems including Unit Test Algorithm Platform http://www.unittestpc.com.cn, Automatic Structural Equation Modeling System http://www.autosem.net, Evolutionary Algorithm Time Complexity Analysis System http://www.eatimecomplexity.net, and Energy Storage Optimization System http://energystorage.autosem. net, which have provided free technical service and support for lots of researchers and engineers.

Affiliations

School of Software Engineering, South China University of Technology, Beijing, P.R. China

Key Laboratory of Big Data and Intelligent Robot (SCUT), MOE of China, Guangzhou, P.R. China

Guangdong Engineering Center for Large Model and GenAI Technology, Guangzhou, P.R. China

Zhifeng Hao

Dr. Hao is a Professor and Doctoral Supervisor of the College of Science, Shantou University. He is currently the Vice Secretary of the Party Committee and President of Shantou University. He received his BS degree in mathematics from Sun Yat-sen University, Guangzhou, China, in 1990, and PhD degree in mathematics from Nanjing University, Nanjing, China, in 1995. He began to work in July 1995 and was promoted to Professor in May 2000. He served as the Dean of the School of Science of SCUT from March 2002, Deputy Director of the Management Committee of SCUT South Campus from May 2003, Executive Deputy Director of the Management Committee of SCUT South Campus from January 2008, member of the Standing Committee of the Party Committee and Vice President of Guangdong University of Technology from November 2008, and Deputy Secretary of the Party Committee and President of Foshan University of Science and Technology from November 2015. He was appointed as the Deputy Secretary of the Party Committee and President of Shantou University in May 2021. At present, he is also the Vice Chairman of the College Mathematics Teaching Steering Committee of the Ministry of Education, the Vice Director of the Big Data and Artificial Intelligence Committee of the Chinese Society of Industrial and Applied Mathematics, the member of the Organizing Committee of the National College Student Mathematical Contest in Modeling, and the Chairman of the Guangdong Supercomputer Application Industry Alliance. His main research interests include algebra and its applications, data science theory, artificial intelligence, and mathematical modeling. He has presided over more than 40 national, provincial, and ministerial projects such as the National Key

Research and Development Program and the National Natural Science Foundation–Guangdong Joint Fund Project. In recent years, he has published more than 80 papers in important domestic and international journals, such as *TNNLS*, *TKDE*, *PR*, *Bioinformatics*, and *Science China*. He has been awarded the second prize of the National Teaching Achievement Award (2005, 2009, and 2018), the Guangdong Science and Technology Awards (first prize for 2016 and second prize for 2014), the second prize of the Natural Science Award of Guangdong Province (2005), the second prize of the Natural Science Award of the Ministry of Education (2002), and the first prize of the Teaching Achievement Award of Guangdong Province (2001, 2005, 2009, 2014, and 2018), the 10th Guangdong Youth May Fourth Medal, the 9th Guangdong Dingying Science and Technology Award, the China Patent Excellence Award (2019), "Leader of the 2020 Smart City Pioneer List," the New Century Excellent Talents Support Plan, and "Guangdong Excellent Teacher" of Guangdong Province.

Affiliation

College of Science, Shantou University, Guangdong, P.R. China

Preface

Intelligent algorithms are the algorithmic forms of computational intelligence methods. These algorithms are the common technical methods used by artificial intelligence in intelligent perception, intelligent decision-making, and intelligent planning. At this stage, intelligent algorithms are usually manifested as evolutionary computation, swarm intelligence, machine learning, other algorithms, and their combinations. Existing literature on intelligent algorithms focuses on theories, methods, and simulation experiments, leaving practical applications underexplored, especially the application of cutting-edge research in artificial intelligence. This book aims to serve as a reference book with practical cases for practitioners of intelligent algorithms based on cutting-edge research results.

In the field of computer vision, the application of deep learning methods is more common; however, there are few reports on the application of intelligent optimization algorithms and their combination with deep learning methods. The first chapter of this book starts with a basic technical problem of computer vision, that is, image matting, and introduces research cases of intelligent algorithms from mathematical modeling and algorithmic design to engineering applications. Image matting is the key technology and basic method of image processing and video analysis. The breakthrough in accuracy and speed will greatly improve the technical performance of subsequent applications. This book introduces a variety of intelligent algorithms, including fuzzy multiobjective evolution algorithms. It also introduces the application of image matting algorithms in the preprocessing of deep learning training data (face detection, face recognition, pedestrian detection, and so on), which can provide implications for technology research and development in computer vision.

Software testing is a necessary but laborious technical activity of software development in the field of software engineering. The technology of software test case generation can save considerable costs of manual operation in program unit testing. However, the test case coding space of actual software programs is often large, resulting in high computational costs of test case generation technology. Based on the idea of heuristic optimization, this book introduces intelligent algorithm strategies such as adaptive evaluation function and association matrix. These strategies realize the optimal allocation of intelligent algorithm computational resources, thus significantly improving the performance of automated software test case generation algorithm. The actual effect of intelligent algorithms has passed the experimental verification of actual software toolkits and public programs such as fog computing, natural language processing, and blockchain smart contracts, thus providing a reference for the research and development of intelligent optimization software engineering technology.

Intelligent logistics is a hotspot topic in the research on artificial intelligence and modern logistics. The two-echelon vehicle routing problem (2E-VRP) is a key optimization problem in the logistics system in the new era. It is also an open problem in the field of management science. This book will explain the background, mathematical model, and research status of 2E-VRP, focus on the cutting-edge intelligent algorithms for this problem, and highlight the new technology of using fuzzy evolution algorithms to solve the contradiction of two-level scheduling, so as to achieve high-precision solution of 2E-VRP. This book also shows the application and related industrialization examples of these research results in urban logistics planning and large-scale port business scheduling, which can provide a reference for the management planning and technology development of intelligent logistics.

In addition to a few applications, this book introduces the latest application of multiobjective optimization intelligent algorithms in software configuration, hoping to provide implications for researchers and technicians who use multiobjective optimization algorithms to solve practical complex optimization problems. The last chapter introduces the computation time estimation method that bridges the gap between theoretical research and practical application, offering theoretical support for the application of intelligent algorithms and providing a practical tool for time complexity analysis.

This book aims to start further discussion on the theory and practice of intelligent algorithms, hoping to provide reference information for practitioners engaged in the application of intelligent algorithms.

Han Huang

Introduction

This book mainly introduces research cases about the theory and practice of intelligent algorithms represented by intelligent optimization algorithms and deep learning methods. These cases cover the application of computer vision, logistics planning, natural language processing, software engineering, and other popular fields. This book also introduces the latest application of multiobjective optimization intelligent algorithms in software configuration and explains in detail the computation-time estimation method that connects theoretical research and practical application, thus providing theoretical support for the application of intelligent algorithms. This book can be used as a textbook for advanced algorithm courses in colleges and universities, such as computer application technology, software engineering, and artificial intelligence. It can also be used for optional reading for algorithm engineers.

CHAPTER 1

Application of intelligent algorithms in the field of computer vision

1.1 Precise capture of translucent visual effects: sampling algorithm based on pixel-level multiobjective optimization

Image matting technology originated from image composition is a type of image processing technology used for accurately extracting the foreground image. Sampling-based image matting algorithms are currently the mainstream matting technology. This technology aims to find the optimal foreground and background pixel pair (referred to as a pixel pair in the following text) for each unknown region pixel, achieving the estimation of alpha mattes for image matting and obtaining the corresponding foreground color.

1.1.1 Overview of Research Progress

In sampling-based image matting algorithms, pixel sampling is a crucial step in reducing the decision space for the optimization problem of foreground/background pixel pairs. As evaluating all possible foreground/background pixel pairs is impossible due to the large size of the decision space, pixel sampling provides a selection of candidate foreground and background pixel samples for optimizing the pairs. The accuracy of the optimization is bound by the quality of the pixel sampling, as the optimal sample may not have been collected. Fig. 1.1 illustrates that pixel sampling is an important step between the preprocessing and the optimal pixel-pair selection in sampling-based image matting algorithms. Considering that evaluating all possible foreground/background pixel pairs is impractical, sampling is necessary for reducing the number of feasible solutions,

Figure 1.1 Sampling-based image matting algorithm flowchart.

Intelligent Algorithms
DOI: https://doi.org/10.1016/B978-0-443-21758-6.00002-4

making optimal pixel pair selection feasible, and selecting the optimal foreground/background pixel pairs for each unknown pixel from the samples collected.

Early sampling algorithms mainly achieved local sampling through a single sampling strategy based on spatial distance features [1−3]. Single-feature sampling algorithms determine whether to collect a pixel as a sample based on the similarity between known and unknown region pixels. Because the optimal foreground/background pixel pairs may not fall in the edge region close to known pixels, a single-feature sampling strategy may lead to the loss of optimal samples. To solve this problem, researchers introduced global sampling algorithms that consider multiple image features to ensure the diversity of sampled pixels, such as the clustering-based sampling algorithm [4] and image matting algorithms based on color and texture features [5]. In addition to spatial distance features, multifeature sampling algorithms also consider other sampling features, such as color features [4,6−8] and texture features [5], which effectively improve the diversity of samples. Existing multifeature sampling strategies mainly fuse multiple sampling features in two ways: the first way is to design an objective function with multiple feature terms [4], where each data term corresponds to a sampling feature, and the second way is to concatenate multiple feature vectors into a total feature vector [6−8].

Recent studies have introduced more effective sample diversity preservation strategies, such as sparse coding-based matting algorithms [8] and Kullback-Leibler (KL)-divergence-based sampling algorithms [7]. These strategies provide quantitative calculation methods to evaluate the representativeness of samples, making it possible to maintain a smaller size of the sample set while ensuring sampling quality, thereby reducing the computational complexity of optimal pixel pair selection. However, these diversity preservation strategies have high time and space complexities, making them inapplicable in the complete sampling space. Since the average color of superpixels[1] is considered to represent the color of the most pixels in the superpixel, these algorithms cluster pixels into superpixels and use the collection of average colors of superpixels as the sampling space [4−8], thereby achieving the goal of reducing the sampling space and completing sample collection within a reasonable time. This compression sampling space strategy is referred to as superpixel-level sampling.

The existing sampling strategies still suffer from the problem of losing the optimal foreground/background pixel pairs, and there are two reasons for this.

1. Existing multi-feature sampling algorithms fail to consider potential conflicts among multiple sampling criteria. They simply merge multiple sampling features through a target function containing multiple data items or through the concatenation of multiple feature vectors, assuming that the sampling criteria for multiple sampling features can be satisfied simultaneously. However, such an assumption may not hold in real sampling scenarios. Existing sampling strategies may miss the optimal foreground/background pixel pairs when there are conflicts among multiple sampling criteria. For example, the sampling criteria for spatial proximity and color similarity may conflict. They cannot be satisfied simultaneously when the optimal foreground/background pixel and the unknown pixel have similar colors but are far apart. As a result, the score of the optimal foreground/background pixel in satisfying the criterion for spatial proximity may not be high compared to the known area pixels that satisfy such criterion, leading to the omission of the optimal foreground/background pixel pairs in existing sampling algorithms.

2. Although superpixel-level sampling can significantly reduce the time and space complexities of sampling algorithms, it makes the sampling space incomplete. The incomplete sampling space may not include the optimal foreground/background pixel pairs. The average color of a superpixel often cannot represent the color of outliers in the superpixel, and these outliers may be the optimal foreground/background pixels. Fig. 1.2 shows an example of the loss of optimal foreground/background pixel pairs caused by superpixel-level sampling.

It can be seen from Fig. 1.2C that there is a significant difference between the average color of the superpixels and the pixel colors marked by the triangles inside the superpixels, while the pixels marked by the triangles are the optimal foreground pixels of the unknown pixels with cross marks in Fig. 1.2B. The loss of optimal foreground/background pixels leads to significant errors in the estimation of alpha mattes (as shown in Fig. 1.2B). This example illustrates that the average color of superpixels often cannot represent the colors of all pixels within the superpixels.

1.1.2 Scientific principles

1.1.2.1 Problem description

This section introduces the mathematical model of pixel sampling involved in optimizing foreground/background pixel pairs in natural image matting. Considering that pixel sampling has a huge number of

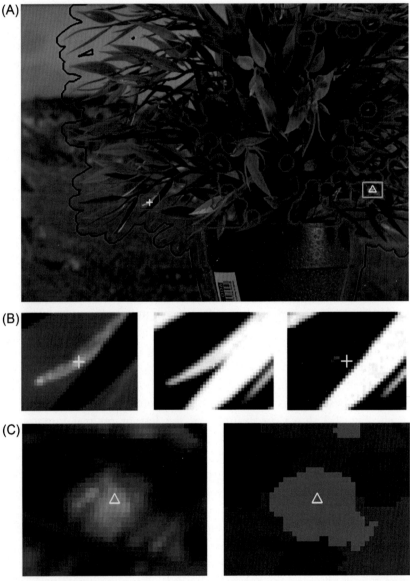

Figure 1.2 Example of loss of optimal foreground/background pixel pairs caused by superpixel-level sampling. (A) The input image. The boundaries of the known foreground region and known background region in the trimap are marked in red and blue, respectively. (B) From top to bottom: a locally enlarged image of the purple region in (A), a ground-truth alpha matte of the locally enlarged purple region in (A), and an alpha matte obtained using a KL-divergence-based sampling matting algorithms [7] of the locally enlarged purple region. (C) From top to bottom: a locally enlarged image of the green region in (A), a locally enlarged superpixel image of the green region in (A), where the pixel colors in the superpixels are replaced with the average color of the superpixel. *KL*, Kullback-Leibler.

possible foreground/background pixel combinations, it is impossible to evaluate all possible foreground/background pixel pairs in an effective time frame. Thus, the selection of the optimal foreground/background pixel pairs cannot be achieved. Therefore, pixel sampling is the process of obtaining a smaller sample set of foreground/background pixels through a selection strategy from the set of known foreground/background pixels in a given trimap.

Let Ω_F and Ω_B be the given pixel sets of known foreground and background regions in a given trimap. $P(*)$ denotes a pixel sampling algorithm P for sampling pixel sets $*$. The known foreground and background pixel sampling problem can be modeled as follows:

$$SF = P(\Omega_F) \; s.t. |SF| \ll |\Omega_F|, \tag{1.1}$$

$$SB = P(\Omega_B) \; s.t. |SB| \ll |\Omega_B|. \tag{1.2}$$

The foreground and background pixel sample sets obtained from the collection are represented by SF and SB, respectively, where $|*|$ denotes the cardinality of the set $*$. A high-quality pixel sample set should satisfy the following conditions for any unknown pixel z:

$$\exists p \in SF, \exists q \in SB, |\hat{\alpha}_{z^{p,q}} - \tilde{\alpha}_z| < \varepsilon \tag{1.3}$$

$$\hat{\alpha} = \frac{(I - B) \cdot (F - B)}{\|F - B\|^2} \tag{1.4}$$

where $\hat{\alpha}z^{p,q}$ represents the alpha value of pixel z estimated by substituting the corresponding color values of the foreground/background pixel pair (p,q) into Eq. (1.4), $\tilde{\alpha}_z$ is the alpha value of pixel z *in* the ground-truth alpha matte, and ε is a small numerical constant.

When the above conditions are met, each unknown pixel has pixel pairs in the sample set that make its estimated alpha value close to the value in the ground-truth alpha matte. When the above conditions are not met, for some or all unknown pixels, all possible foreground/background pixel pairs formed by foreground and background pixels in the sample set have significant differences compared to the optimal foreground/background pixel pairs corresponding to the unknown pixel, causing significant deviation between the estimated alpha value and the value in the ground-truth alpha matte, thus leading to the problem of losing the optimal foreground/background pixel pairs.

1.1.2.2 A multiobjective image matting algorithm based on pixel-level global sampling

To solve the problem of the slow speed in optimization-based image matting algorithms, a pixel-level multiobjective global sampling-based image matting algorithm [1] is proposed based on the combination of optimization-based image matting algorithms and sampling-based image matting algorithms. The core of this algorithm is the pixel-level multiobjective global sampling algorithm. Unlike existing sampling algorithms that assume that the optimal foreground/background pixel pairs are distributed in different specific regions, the introduced sampling algorithm uses multiobjective optimization to achieve pixel sampling. Compared with existing sampling algorithms, the pixel-level multiobjective global sampling algorithm has two main innovations.

1. The adaptive trade-off between multiple sampling criteria is achieved through multiobjective optimization, which resolves conflicts between sampling criteria corresponding to different sampling features. For an unknown pixel, the problem of sampling foreground/background pixels from multiple sampling features is modeled as a discrete multiobjective optimization problem. The introduced algorithm solves the discrete multiobjective optimization problem and samples all pixels in the Pareto optimal[1] set as global samples. Because Pareto optimal solutions are independent of the scales of different sampling objectives, the algorithm does not require weighting of different sampling criteria and does not involve tuning empirical parameters.

2. Different from superpixel-level sampling which reduces sampling space by clustering pixels into superpixels, the algorithm introduced in this section collects samples in the entire sampling space, avoiding the problem of losing the optimal foreground/background pixel pairs due to incomplete sampling space. Therefore, a strategy called pixel-level sampling involves collecting every pixel in the known foreground/background regions as a pixel sample.

In the pixel-level multiobjective global sampling algorithm, three sampling criteria were designed for the evaluation function of foreground/background pixel pairs, including similarity in color with unknown pixels, proximity in spatial distance, and similarity in texture. The problem of the

---

[1] Pareto optimal solution refers to the feasible solution in which no other better feasible solutions exist in each objective. Pareto optimal solution is the best solution for multiobjective optimization problems.

foreground/background pixel sampling for an unknown pixel is modeled as a discrete multiobjective optimization problem with two or more targets. To quickly solve a large number of multiobjective optimization problems, this section introduces a fast discrete multiobjective optimization (FDMO) algorithm[2] with a time complexity of $O(kn)$ and a space complexity of $O(n)$. All pixels in the Pareto optimal solution set are collected as samples.

The pixel-level multiobjective global sampling-based image matting algorithm uses an optimal pixel pair selection method based on a pixel-objective function to obtain the corresponding alpha value. Specifically, the foreground sample set and the background sample set obtained by the pixel-level multiobjective global sampling algorithm generate foreground/background pixel pair candidates through the Cartesian product. By minimizing a foreground/background pixel pair objective function that includes two widely used evaluation items [3−8], the foreground/background pixel pair with the optimal objective function value is selected from the pixel pair candidate set, and its corresponding color is substituted into Eq. (1.4) to estimate the alpha value of unknown pixels. The foreground/background pixel pair objective function used includes a color distortion term and a spatial distance term. Given an unknown pixel z and a foreground/background pixel pair (x_F, x_B) in the candidate set, where $x_F \in SF$ and $x_B \in SB$, the foreground/background pixel pair objective function can be expressed as follows:

$$O_z(x_F, x_B) = O_z^{(c)}(x_F, x_B) \times O_z^{(s)}(x_F, x_B) \tag{1.5}$$

The definitions of the color distortion term $O_z^{(c)}(x_F, x_B)$ and the spatial distance term $O_z^{(s)}(x_F, x_B)$ are given by the following two formulae:

$$O_z^{(c)}(x_F, x_B) = \exp\left(-\left\| C_z - \left(\hat{\alpha}_z C_{x_F} + (1 - \hat{\alpha}_z)C_{x_B}\right)\right\|\right) \tag{1.6}$$

$$O_z^{(S)}(x_F, x_B) = \exp\left(\frac{-\left\|S_z - S_{x_F}\right\|}{\frac{1}{|\Omega_F|}\sum_{x_i \in \Omega_F}\left\|S_z - S_{x_i}\right\|}\right) \times \exp\left(\frac{-\left\|S_z - S_{x_B}\right\|}{\frac{1}{|\Omega_B|}\sum_{x_j \in \Omega_B}\left\|S_z - S_{x_j}\right\|}\right) \tag{1.7}$$

The following will introduce the pixel-level discrete multiobjective sampling (PDMS) strategy and the FDMO algorithm involved in the pixel-level multiobjective global sampling algorithm.

1.1.2.2.1 Pixel-level discrete multiobjective sampling strategy

To solve the conflicting problem among multiple sampling criteria for different sampling features, this section introduces the PDMS strategy. This strategy models the foreground/background pixel sampling problem for each unknown pixel as a discrete multiobjective optimization problem and defines sampling criteria that can approximate the optimal solution for the evaluation function, with each sampling criterion modeled as an optimization objective. As described in Section 1.1.1, an incomplete sampling space can lead to the loss of the optimal foreground/background pixel pairs. To avoid this problem, the PDMS strategy samples from the set composed of all known pixels in the region, achieving pixel-level sampling, which is different from the superpixel-level sampling that samples from a set composed of the average color of superpixels.

This section first introduces the sampling criteria involved in the PDMS strategy. Liang et al. pointed out that sampling algorithms based on a single spatial proximity sampling criterion can lead to the problem of losing the optimal sample [4]. To solve this problem, the PDMS strategy adopts three sampling features, i.e., color, space, and texture. It defines separate sampling criteria for each of these features.

Color is an important feature of image matting. The PDMS strategy adopts an effective sampling criterion that selects pixels with colors that are similar to the given unknown pixel z. Given an unknown pixel z and a pixel x from a known (foreground /background) region, the color similarity sampling criterion can be modeled as the following objective function:

$$g_1(x) = \left\| C_x - C_z \right\|, \tag{1.8}$$

where C_x and C_z respectively represent RGB color space vectors of the known region pixel x and unknown region pixel z, and $\| * \|$ represents the magnitude of vector $*$.

Considering the widely used spatial distance feature, the PDMS strategy adopts the sampling criterion of being close to the given unknown pixel spatial distance. The sampling criterion of spatial distance proximity can be modeled as the following objective function:

$$g_2(x) = \left\| S_x - S_z \right\|, \tag{1.9}$$

where S_x and S_z represent the spatial coordinate vectors of the known region pixel x and unknown region pixel z, respectively.

In addition, Shahrian et al.'s research shows that texture features can effectively improve pixel sampling performance in complex scenes for image matting [5]. The PDMS strategy also incorporates the texture features of the image and establishes a texture similarity sampling criterion based on the local binary pattern. Its corresponding objective function can be expressed as follows:

$$g_3(x) = \left\| T_x - T_z \right\|,\qquad(1.10)$$

where T_x and T_z represent the local binary pattern feature vectors of pixel x in the known region and those of pixel z in the unknown region, respectively.

As mentioned at the beginning of this section, the purpose of multiobjective sampling is to solve the problem of losing the optimal foreground/background pixel pairs due to conflicting sampling criteria. In multifeature-based sampling algorithms, the sampling criteria corresponding to multiple features often cannot be satisfied simultaneously. For example, the optimal foreground/background pixel may have a similar color to the given unknown pixel but be located far away from the unknown pixel. In this case, although the color difference between known region pixels and unknown pixels is small, the spatial distance between them is large, resulting in conflicts between the color similarity sampling criterion and the spatial proximity sampling criterion, which means that the two criteria cannot be satisfied simultaneously.

The PDMS strategy models foreground and background pixel sampling as two discrete multiobjective optimization problems. The multiobjective optimization problems aim to optimize all objective functions simultaneously. If there are n sampling criteria, the multiobjective optimization problem corresponding to the foreground pixel sampling problem can be expressed as follows:

$$\min g_1(x), \min g_2(x), \ldots, \min g_n(x) \quad \text{s.t.} \quad x \in \Omega_F,\qquad(1.11)$$

where the function $g_i(x)$ represents the ith target to be optimized, $i = 1,2,\ldots,n$, the decision variable x is the one-dimensional index of known foreground region pixels, $x = 1,2,\ldots,|\Omega_F|$. Since the decision variable takes discrete integer values, this problem is a discrete multiobjective optimization problem. Conflicts often exist among the objectives involved in multiobjective optimization problems, and no solution can make all objectives optimal simultaneously. Therefore, the optimal solution to multiobjective optimization problems refers to the Pareto optimal solution, which is a feasible solution where there is no other solution better than it on each target. The PDMS strategy takes all Pareto optimal solutions of the multiobjective optimization problem as pixel samples. A similar approach can be used to establish a multiobjective optimization model for background pixel sampling problems.

To avoid the problem of losing the optimal foreground/background pixel pairs due to incomplete sampling space, we introduce a pixel-level sampling strategy to collect pixel samples from the set of all pixels in the known foreground/background region. To obtain a complete sampling space, the PDMS strategy does not use the pixel clustering step commonly used in superpixel-level sampling algorithms [4−8]. Therefore, every pixel in the known foreground/background region may be selected as a sample. The diversity of pixel samples is a key factor in obtaining high-quality segmentation results [4−6]. Compared to superpixel-level sampling, the PDMS strategy increases sample diversity by sampling in a complete sampling space.

1.1.2.2.2 Fast discrete multiobjective optimization algorithm

The PDMS strategy models the foreground/background sampling problem of each unknown pixel as a discrete multiobjective optimization problem, which brings about a huge number of multiobjective sampling optimization problems. Effectively addressing the significant challenge of solving large quantities of discrete multiobjective optimization problems in a limited time is a key issue faced by global sampling algorithms for PDMS. Although the discrete multiobjective optimization problem is a combinatorial optimization problem, considering the factors such as the small number of objectives involved in the discrete multiobjective optimization problem, low dimensionality of decision variables, and not particularly large number of feasible solutions, the single multiobjective optimization problem involved in the PDMS strategy is not difficult to solve. One approach to solving the problem is to use a brute force method with a time complexity of $O(n^2)$. On the one hand, due to the pixel-level sampling strategy, the number of feasible solutions for the multiobjective optimization problem is greatly increased compared to the superpixel-level sampling. On the other hand, for an image with 300,000 pixels, the number of multiobjective optimization problems involved in pixel-level discrete sampling can reach over 10,000. Therefore, using the brute force calculation is very time-consuming and infeasible in practical applications.

The existing algorithms for solving multiobjective optimization problems can be divided into two categories: deterministic algorithms and approximation algorithms. Deterministic algorithms, exemplified by the skyline algorithm [9], can ensure the exact solution of all Pareto optimal solutions. Such algorithms can solve multiobjective optimization problems for which feasible solution numbers increase linearly with problem size. Since these algorithms are designed for Structured Query Language (SQL) queries, they are mainly used to solve small numbers of multiobjective optimization problems with a relatively large number of feasible solutions. To implement SQL queries in a large number of records, these algorithms are designed to consider characteristics such as the fact that all feasible solutions will only be accessed once. These characteristics are certainly important for SQL queries that involve billions of records, but they do not offer significant advantages for multiobjective optimization problems involved in pixel-level discrete sampling strategies, because such problems only involve a relatively small number of feasible solutions. Moreover, implementing these characteristics may incur unnecessary computational overhead resulting in longer sampling times. Approximate algorithms, represented by heuristic optimization algorithms, aim to obtain approximate solutions to complex multiobjective optimization problems that deterministic algorithms are unable to solve due to their NP-hard complexity. The drawback of approximation algorithms is that they have high time complexity and cannot solve large-scale multiobjective optimization problems within an effective time frame.

To efficiently solve the large-scale discrete multiobjective optimization problems involved in the pixel-level sampling strategy, we introduce the FDMO algorithm[3] with a time complexity of $O(kn)$ and a space complexity of $O(n)$. The FDMO algorithm is a deterministic algorithm, which guarantees to find all Pareto optimal solutions of the multiobjective optimization problem. The basic idea of the algorithm is to find a Pareto optimal solution in each iteration and remove the dominated solutions from the candidate set in the comparison process of the algorithm iteration, thereby avoiding unnecessary computational complexity.

---

[3] The variable n represents the number of feasible solutions in a discrete multiobjective optimization problem, while k represents the number of Pareto optimal solutions.

The four steps of the FDMO algorithm include the following:

1) Choose a feasible solution from the feasible solution set Ω; for the foreground pixel sampling problem, $\Omega = \Omega_F$; for the background pixel sampling problem, $\Omega = \Omega_B$.

2) Compare the selected feasible solution with the other solutions in the feasible solution set Ω^4 one by one. If one solution dominates another solution during the comparison process, the dominated solution is removed from Ω. If the selected solution is removed, another solution participating in the comparison becomes the selected solution, and Step 2 is repeated.

3) Add the selected feasible solutions to the Pareto optimal solution set and remove them from Ω.

4) Repeat the above steps until Ω is empty.

Algorithm 1.1 provides the specific implementation of the FDMO algorithm, where swap() represents an exchange operator. After each round of iteration, this algorithm can ensure that a Pareto optimal solution is found. Suppose that a given discrete multiobjective sampling problem contains k Pareto optimal solutions. The FDMO algorithm can find all Pareto optimal solutions in k rounds of iterations. In the comparison process, non–Pareto optimal solutions are removed from the Ω set. The number of comparisons required per iteration will gradually decrease as the cardinality of the Ω set continuously decreases. For multiobjective sampling problems, the number of Pareto optimal solutions is usually smaller than the number of feasible solutions. Taking the foreground multiobjective pixel sampling problem of a plant image in the image matting benchmark test set [10] as an example, there are 168,409 feasible solutions for the corresponding multiobjective pixel sampling problem of this image, of which only 11 feasible solutions are Pareto optimal solutions. In multiobjective pixel sampling problems, the number of Pareto optimal solutions is much smaller than the number of feasible solutions. Therefore, a large number of dominated feasible solutions will be removed in each round of iteration of the FDMO algorithm, and the number of comparisons required in the next round of iteration will be greatly reduced, thereby achieving fast solving of multiobjective optimization problems.

---

[4] In multiobjective optimization problems, when a solution x is not inferior to another solution y in all dominates and is better than y in at least one objective, x dominates y [10].

ALGORITHM 1.1 Pseudocode of fast discrete multiobjective optimization algorithm for pixel sampling in image matting

Input: A discrete multiobjective foreground/background sampling problem, an array C containing all feasible solutions to the problem

output: Modified array C of feasible solutions, i (The first $i - 1$ solutions of the array C are Pareto optimal solutions)

```
 1: i ← 1
 2: j ← Length of array C
 3: while i ≤ j do
 4:     cmp ← i + 1
 5:     while cmp ≤ j do
 6:         if The ith feasible solution in the array C dominates the cmp-th
            feasible solution then
 7:             swap(C[cmp], C[j])
 8:             j ← j - 1
 9:         else if The cmp-th feasible solution in the array C dominates the ith
            feasible solution then
10:             swap(C[i], C[cmp])
11:             swap(C[cmp], C[j])
12:             j ← j - 1
13:             cmp ← i + 1
14:         else
15:             cmp ← cmp + 1
16:         end if
17:     end while
18:     i ← i + 1
19: end while
```

The following is an analysis of the time and space complexities of the introduced FDMO algorithm. In the complexity analysis, we consider the comparison of domination relationships between two feasible solutions as the basic operation of the algorithm, and we also consider both the time and space complexities of this operation as $O(1)$. First, we analyze the time complexity of the algorithm in the best, worst, and average cases. Assume that the feasible set of the given multiobjective sampling problem is Ω and contains k Pareto optimal solutions. The FDMO algorithm can find a Pareto optimal solution in each iteration, and all Pareto optimal solutions can be obtained after k iterations, satisfying the stopping condition. Since at least one feasible solution is removed from the set Ω in each iteration, the number of comparisons required for the ith iteration is no more than $2(n - 1 - i)$ times. After k iterations, the computational complexity of the FDMO algorithm can be known as $O(kn)$. In the best case, there is only one feasible solution in the entire feasible set that is Pareto optimal. That is, $k = 1$. All Pareto optimal solutions can be obtained in the first iteration, and since all other feasible solutions are dominated by this Pareto optimal solution, they are all removed from the Ω set in the comparison. After the first iteration, the Ω set is an empty set, and the stopping condition is met. Therefore, in the best case, the time complexity of the FDMO algorithm is $O(n)$. In the worst case, all feasible solutions are Pareto optimal. That is, k is equal to n. Only the Pareto optimal solution obtained in each iteration is removed from the Ω. Therefore, the computational complexity in the worst case is $O((1 + n)n/2)$, which is simplified to $O(n^2)$. Since the FDMO algorithm does not need to occupy any other storage space except for the array containing all feasible solutions and limited temporary variables in all cases, its space complexity in the best case, worst case, and average case is $O(n)$. Table 1.1 summarizes the time and space complexity analysis results of the FDMO algorithm.

1.1.2.3 Experimental results and discussion

Two experiments are conducted to comprehensively test the introduced global pixel-level multiobjective sampling algorithm. The first experiment verifies whether the sampling algorithm solves the problem of losing the optimal foreground/background pixel pairs due to conflicting sampling criteria or incomplete sampling space. The second experiment verifies whether the introduced sampling algorithm can improve the quality of alpha mattes obtained by sampling-based image matting algorithms. The images used in the experiment come from Rhemann et al.'s benchmark dataset for objective evaluation of image matting alpha mattes [10]. The dataset contains 35 natural images, of which 27 are training images with ground-truth alpha mattes provided, and the remaining 8 are test images without publicly available image matting alpha mattes. The dataset provides three different types of trimap for the images in the test set, one

Table 1.1 Time and space complexity analysis results of the FDMO algorithm.

	The best case	The worse case	Average
Time complexity	$O(n)$	$O(n^2)$	$O(kn)$
Space complexity	$O(n)$	$O(n)$	$O(n)$

manually labeled by users and the other two automatically generated by morphological operations on the ground-truth alpha mattes using different sizes of structuring elements (the trimap obtained with smaller structuring elements has a smaller unknown area, while the one obtained with larger structuring elements has a larger unknown area). The dataset only provides automatically generated trimaps for the images in the training set. The experiment is conducted on a server with dual Intel Xeon E5 2620 central processors and 32 GB of memory.

1.1.2.3.1 Performance verification experiment of the pixel-level multiobjective global sampling algorithm

This experiment verifies the sampling performance of the pixel-level multiobjective global sampling algorithm by comparing it with two advanced image pixel sampling algorithms. In this experiment, we use two evaluation metrics to quantitatively evaluate the sampling performance of the introduced pixel-level multiobjective global sampling algorithm. The first evaluation metric is the minimum absolute error of the alpha value. The sum of the absolute errors has been widely used for the objective evaluation of alpha mattes in image matting [3–8,10]. In sampling-based image matting algorithms, given a foreground/background pixel pair, the alpha value of an unknown pixel can be obtained through Eq. (1.4). That is, for an unknown pixel, the foreground/background pixel pair directly determines its estimated alpha value. Therefore, the quality of foreground/background pixel pairs can be evaluated by the error of the alpha matte. Given an unknown pixel, compared to other foreground/background pixel pairs, if a pair has an alpha value that is closer to the alpha value in the ground-truth alpha matte, the pair is considered better than other pixel pairs. This experiment compares the performance of different pixel sampling algorithms by comparing the absolute alpha value error corresponding to foreground/background pixel pairs. Firstly, the foreground/background pixel pair candidate set is obtained by combining the foreground sample set with the background sample set through sampling. Then the alpha value corresponding to each foreground/background pixel pair is calculated and compared with the corresponding alpha value in the ground-truth alpha matte. The minimum absolute error of the alpha value in the candidate set is taken as the evaluation metric of the sampling algorithm. The smaller the minimum absolute error of the alpha value, the fewer situations where the optimal foreground/background pixel pairs are lost, indicating better sampling performance. The corresponding transparencies of the pixel pairs in the candidate set have significant deviations from those in the ground-truth alpha matte when the optimal foreground/background pixel pairs are lost, leading to a significant increase in the value of the minimum absolute error of the alpha value metric. The second evaluation metric used in this experiment is the cardinality of the candidate pixel pair set generated by taking the Cartesian product of the foreground and background sample sets obtained by sampling. Recent studies have shown that a small number of pixel samples can effectively represent the colors of unknown areas [7,8]. Since the selection of the optimal pixel pair will evaluate the pixel pairs in the candidate set one by one, a large candidate set will consume a lot of time. If two sampling algorithms have the same minimum absolute error of the alpha value for the pixel sample sets they obtained, and one algorithm generates a large candidate set, the sampling algorithm that generates a smaller candidate set has better practicality. The cardinality of the candidate pixel pair set is of great significance for the practical application of image matting.

In this experiment, a benchmark dataset of 27 images with standard alpha mattes for foreground extraction and corresponding automatically generated trimaps with small unknown regions were used. The 27 images used contain 38,573, 57,461, 121,014, 178,728, 29,357, 55,831, 52,970, 142,232, 90,208, 51,540, 61,930, 33,836, 143,935, 37,039, 43,538, 87,834, 51,098, 56,751, 25,462, 49,105, 121,177, 65,605, 63,437, 49,174, 59,459, 153,224, and 102,471 unknown pixels.

Two advanced pixel sampling algorithms were used as benchmarks for sampling performance in this experiment, i.e., the KL-divergence-based sampling algorithm [7] and the comprehensive sampling algorithm [6]. The KL-divergence-based sampling algorithm collected all pixels sampled from the GT02 and GT25 images within the known foreground regions, but failed to collect any background pixels, resulting in inconclusive experimental results.

Fig. 1.3A shows the minimum absolute alpha value error of the pixel-level multiobjective global sampling algorithm and two other advanced pixel sampling algorithms on 27 images in the form of box plots. Each column in this figure describes the distribution of the minimum absolute alpha value error of a sampling algorithm for all unknown pixels in a test dataset. As shown in Fig. 1.3A, the pixel-level multiobjective global sampling algorithm has a lower minimum absolute alpha value error in most of the images compared with existing pixel sampling algorithms, indicating that the sample set obtained by the introduced algorithm for different

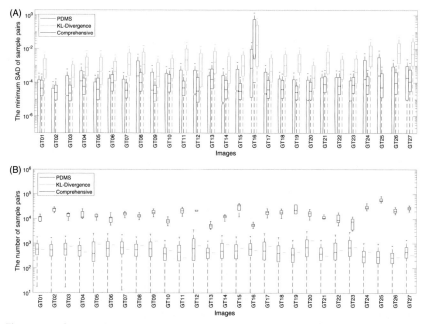

Figure 1.3 Quantitative comparison experiment results between pixel-level multiobjective global sampling algorithm and advanced pixel sampling algorithm. (A) Comparison of minimum transparency absolute error of foreground and background pixel pairs obtained by the sampling algorithm. (B) Comparison of the cardinality of candidate foreground and background pixel pairs obtained by the sampling algorithm.

unknown pixels contains foreground/background pixel pairs with small alpha value errors. In image GT16, the sampling algorithms based on KL divergence and combined sampling have a minimum absolute alpha value error exceeding 0.5 for more than a quarter of the unknown pixels, indicating the problem of losing optimal foreground/background pixel pairs during the sampling process. The quartile range of the minimum absolute alpha value error corresponding to the pixel-level multiobjective global sampling algorithm is only about 0.1, which is much lower than the other two sampling algorithms. This experimental result shows that the pixel-level multiobjective global sampling algorithm effectively alleviates the problem of losing the optimal foreground/background pixel pairs. Fig. 1.3B shows the box plots of the size of the candidate pixel pair set corresponding to the pixel-level multiobjective global sampling algorithm and the two other advanced pixel sampling algorithms on 27 images.

Each column in this figure describes the distribution of the candidate pixel pair set size for all unknown pixels corresponding to a sampling algorithm in a test dataset. In addition, it can be found from Fig. 1.3B that the introduced pixel-level multiobjective global sampling algorithm and the sampling algorithm based on KL divergence obtain fewer candidate foreground/background pixel pair sets, which are one order of magnitude lower than the candidate pairs obtained by the combined sampling algorithm. This experimental result further illustrates that the pixel-level multiobjective global sampling algorithm can use a smaller candidate set to cover high-quality foreground/background pixel pairs for different unknown pixels, achieving precise pixel sampling.

Tables 1.2 and 1.3 report the first, second, and third quartiles of the minimum alpha value absolute error of three pixel sampling algorithms. The last column of Table 1.3 summarizes the number of images in which the minimum alpha value absolute error index of the pixel-level multiobjective global sampling algorithm in this quartile is superior or not among the 27 images. Compared with the sampling algorithm based on KL divergence, the first, second, and third quartiles of the minimum alpha value error of the pixel-level multiobjective global sampling algorithm are superior in all participating images. Compared with the comprehensive sampling algorithm, the third quartile of the minimum alpha value error of the pixel-level multiobjective global sampling algorithm is superior in all participating images, while its first and second quartiles of the minimum alpha value absolute error are superior in most participating images.

The above experimental results show that the pixel-level multiobjective global sampling algorithm has the advantage of having a small cardinality for the foreground/background pixel candidate set, which can obtain a small absolute alpha value error, and can adapt to different unknown pixel changes in images in various scenes. This experimental result also shows that the pixel-level multiobjective global sampling algorithm accurately covers the optimal foreground/background pixel pairs of most unknown pixels and has good pixel sampling performance.

To further analyze the sampling performance, the GT16 image was taken as an example in this experiment to compare the distribution of pixel samples obtained by the three pixel sampling algorithms. Fig. 1.4 shows the distribution of pixel samples obtained by the pixel-level multiobjective global sampling algorithm and the two other advanced sampling algorithms. " + " symbols denote unknown pixels. Foreground and

Table 1.2 Results of minimum alpha value error quantile comparison between pixel-level multiobjective global sampling algorithm and existing pixel sampling algorithm.

Sampling algorithms (quartile)	GT01	GT02	GT03	GT04	GT05	GT06	GT07	GT08	GT09	GT10	GT11	GT12	GT13	GT14
Comprehensive (first quartile)	+	+	−	+	+	+	+	+	−	+	−	+	+	−
Comprehensive (second quartile)	+	+	+	+	+	+	+	−	+	+	+	+	+	+
Comprehensive (third quartile)	+	+	+	+	+	+	+	+	+	+	+	+	+	+
KL divergence (first quartile)	+	N/A	+	+	+	+	+	+	+	+	+	+	+	+
KL divergence (second quartile)	+	N/A	+	+	+	+	+	+	+	+	+	+	+	+
KL divergence (third quartile)	+	N/A	+	+	+	+	+	+	+	+	+	+	+	+

KL, Kullback–Leibler.

Note: 1. " + " indicates that the pixel-level multiobjective global sampling algorithm has a smaller minimum alpha value error than that of the sampling algorithms involved in the comparison.

2. " − " means that the minimum alpha value error obtained by the pixel-level multiobjective global sampling algorithm is greater than that of the sampling algorithm compared with it.

3. The smaller the absolute error of the minimum alpha value, the better.

Table 1.3 Results of minimum alpha value error quantile comparison between pixel-level multiobjective global sampling algorithm and existing pixel sampling algorithm.

Sampling algorithms (quartile)	GT15	GT16	GT17	GT18	GT19	GT20	GT21	GT22	GT23	GT24	GT25	GT26	GT27	+/-
Comprehensive (first quartile)	−	+	−	+	+	+	+	+	+	−	−	+	+	19/ 8
Comprehensive (second quartile)	+	+	+	+	+	+	+	+	+	−	−	+	+	24/ 3
Comprehensive (third quartile)	+	+	+	+	+	+	+	+	+	+	+	+	+	27/ 0
KL divergence (first quartile)	+	+	+	+	+	+	+	+	+	+	N/A	+	+	25/ 0
KL divergence (second quartile)	+	+	+	+	+	+	+	+	+	+	N/A	+	+	25/ 0
KL divergence (third quartile)	+	+	+	+	+	+	+	+	+	+	N/A	+	+	25/ 0

KL, Kullback–Leibler.

Note: 1. " + " indicates that the pixel-level multiobjective global sampling algorithm introduced has a smaller minimum alpha value error than the sampling algorithms involved in the comparison.

2. " − " means that the minimum alpha value error obtained by the introduced pixel-level multiobjective global sampling algorithm is greater than the sampling algorithm compared with it.

3. The smaller the absolute error of the minimum alpha value, the better.

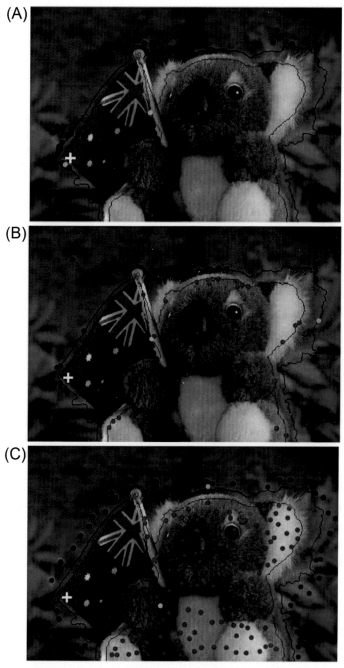

Figure 1.4 Comparison of the pixel sample distributions obtained by the pixel-level multiobjective global sampling algorithm and two other advanced pixel sampling algorithms. (A) PDMS. (B) KL divergence. (C) Comprehensive sampling. *KL*, Kullback-Leibler; *PDMS*, pixel-level discrete multiobjective sampling.

background pixel samples captured by each sampling algorithm are represented by red and blue dots, respectively. The pixel pair with the minimum absolute alpha value error in the candidate set of foreground/background pixel pairs is marked with light red and light blue. In this example, the optimal foreground/background pixel pairs corresponding to unknown pixels make the sampling criteria of similar color and spatial distance conflict. As shown in Fig. 1.4C, although the comprehensive sampling algorithm has collected a large number of samples, it failed to collect foreground samples of the blue flag, resulting in the loss of the optimal foreground/background pixel pairs. As shown in Fig. 1.4A and B, both the pixel-level multiobjective global sampling algorithm and the KL-divergence-based sampling algorithm can collect blue flag samples, and the samples collected by the former are closer to the color of unknown pixels. Its corresponding minimum absolute alpha value error is 0.37, which is much smaller than the minimum absolute alpha value error of 0.89 corresponding to the KL-divergence-based sampling algorithm. This experimental result shows that the pixel-level multiobjective global sampling algorithm can collect more accurate foreground/background pixel pairs than existing sampling algorithms.

This experiment shows that the pixel-level multiobjective global sampling algorithm can adaptively provide a small candidate set of foreground/background pixel pairs but a lower minimum alpha value absolute error for different unknown pixels. The pixel-level multiobjective global sampling algorithm has the advantages of both the sampling algorithm based on KL divergence and the comprehensive sampling algorithm: the former has a smaller base for the foreground/background pixel pair candidate set, and the latter has a smaller corresponding minimum alpha value error for the foreground/background pixel pair candidate set. Even in cases where there are conflicts between multiple sampling criteria, the pixel-level multiobjective global sampling algorithm can still adaptively balance multiple criteria, covering the optimal foreground/background pixel pairs and effectively alleviating the problem of losing the optimal foreground/background pixel pairs.

1.1.2.3.2 Performance verification experiment of the pixel-level multiobjective global sampling-based matting algorithm

This experiment uses the online test of Rhemann et al.'s benchmark dataset [10] for objectively evaluating the image matting algorithm based on pixel-level multiobjective global sampling. The aim is to provide a quantitative analysis of the performance of the pixel-level multiobjective global sampling algorithm.

In this experiment, the gradient error performance indicator was used to quantitatively evaluate the alpha matte obtained from the image matting. The reason for choosing this indicator is that Rhemann et al. pointed out that the gradient error is more comparable to subjective evaluation results than the sum of absolute errors and mean square error [10]. Therefore, using this indicator can obtain more accurate objective evaluation results. The smaller the gradient error, the better the quality of the alpha matte obtained from image matting.

To verify the scalability of the pixel-level multiobjective global sampling algorithm, two versions of the pixel-level multiobjective global sampling-based matting algorithm, namely the dual-objective and triple-objective versions, were tested. The dual-objective version uses similar color and spatial distance as two optimization objectives, while the triple-objective version uses three optimization objectives including similar color, spatial distance, and texture features. To verify the effectiveness of the pixel-level sampling strategy, we replaced the pixel-level sampling strategy in the pixel-level multiobjective global sampling algorithm with the superpixel-level sampling strategy used in existing algorithms, resulting in the superpixel-level discrete multiobjective sampling-based matting algorithm (SDMS). The SDMS was compared with the pixel-level multiobjective global sampling-based matting algorithm in the experiment. In this experiment, the simple linear iterative clustering algorithm proposed by Achanta et al. [11] was used to cluster pixels into superpixels to achieve the superpixel-level sampling strategy. The empirical parameter settings involved were consistent with those in reference [6]. In addition to the three matting algorithms related to the pixel-level multiobjective global sampling algorithm, eight advanced matting algorithms introduced in recent years, namely sparse coding-based matting [8], pixel block-based matting [12], KL-divergence-based sampling matting [7], comprehensive sampling-based matting [6], color and texture feature weighting-based matting [5], K-nearest neighbors (KNN)-based matting [13], global sampling-based matting [3], and sharing-based matting [14], were used as test benchmarks.

The existing image matting algorithms use preprocessing and postprocessing methods to improve the quality of the estimated alpha matte. The preprocessing method expands the known region and reduces the unknown region to obtain more known information, thereby improving the quality of the segmentation. Different preprocessing algorithms are discussed in detail in reference [15]. Sampling-based image matting algorithms do not consider the local smoothness properties among pixels, and postprocessing can make the obtained alpha matte smoother. Considering that existing sampling-based image matting algorithms use both preprocessing and postprocessing to improve the segmentation quality, the pixel-level multiobjective global sampling image matting algorithm and the superpixel-level multiobjective global sampling image matting algorithm use the same preprocessing and postprocessing methods as the comprehensive sampling image matting algorithm to ensure fairness in comparison, and the parameter settings are also consistent with them. The following is a brief introduction to the preprocessing and postprocessing methods used. The preprocessing algorithm expands the known area of the trimap by comparing the similarity between the pixels in the known region and the pixels at the edge of the unknown region. Given a known foreground pixel p, when an unknown region pixel z satisfies the following conditions, the unknown pixel z will be expanded to a known foreground pixel:

$$\text{Condition } 1: \left\| S_p - S_z \right\| < t_{E,} \, p \in \Omega_F$$
$$\text{Condition } 2: \left\| C_p - C_z \right\| < t_{C,} \, p \in \Omega_F \tag{1.12}$$

where C_p and S_p represent the RGB color vector and spatial distance vector of pixel p, respectively, t_C and t_E are the threshold values of color and spatial distance features, respectively. In this experiment, they are all set to 9 as recommended in reference [6]. In the postprocessing of image matting, the estimated alpha matte is obtained by minimizing an objective function containing a Laplacian smoothness term, which can provide a more accurate alpha matte. The following equation gives a mathematical description of the objective function involved in the postprocessing:

$$\alpha = \text{argmin} \alpha^T L \alpha + \lambda (\alpha - \hat{\alpha})^T \Lambda (\alpha - \hat{\alpha}) + \gamma (\alpha - \hat{\alpha})^T \Gamma (\alpha - \hat{\alpha}) \tag{1.13}$$

where λ is a large constant and its value is set to 1000. γ is a parameter that controls the weight between data fidelity and regularization terms, and its value is set to 0.1. Λ and Γ are diagonal matrices. The diagonal elements of Λ corresponding to known pixels are set to 1, and those corresponding to unknown pixels are set to 0. The diagonal elements of Γ corresponding to known pixels are set to 0, and those corresponding to unknown pixels are set to their target function values.

The experiment used three types of trimaps: trimaps with large unknown regions, trimaps with small unknown regions, and manually labeled trimaps. For each type of trimap, the gradient error between the alpha matte obtained by the image matting algorithm in the test set and the ground-truth alpha matte is calculated. The average gradient error of all images is then used as the performance indicator for the corresponding type of trimap. By taking the average of the mean gradient errors corresponding to the three types of trimaps, the overall mean gradient error evaluation index is obtained.

Table 1.4 summarizes the average gradient errors of 11 image matting algorithms on 7 test images of the online image matting benchmark dataset [10] for different types of trimap and the overall average gradient error. As shown in Table 1.4, both versions of the pixel-level multiobjective global sampling image matting algorithm (dual-objective and triple-objective sampling) significantly improved the gradient error of the estimated segmentation mask. The dual-objective and triple-objective sampling versions of the pixel-level multiobjective global sampling image matting algorithm ranked

Table 1.4 Gradient error of image matting algorithms based on pixel-level multiobjective global sampling and existing advanced image matting algorithms on seven test images of the online image matting benchmark.

Image matting algorithms	Overall rank Avg. rank	Avg. small rank	Avg. large rank	Avg. user rank
PDMS-based matting (three objectives)	**11.3**	13.6	**9.8**	**10.4**
PDMS-based matting (two objectives)	12.1	12.5	12.3	11.6
SDMS-based matting (two objectives)	28.7	29.9	29.5	26.9
Graph-based sparse matting [8]	14.8	12	13.6	18.9
Patch-based matting [12]	15.1	**11.3**	14.9	19.1
KL divergence [7]	15.2	13.0	14.4	18.1
Comprehensive sampling [6]	16.6	16.5	17.0	16.3
Weighted color and texture [5]	27.3	25.9	26.8	29.3
KNN matting [13]	31.3	34.4	31.4	28.1
Global sampling matting [3]	25.9	23.1	26.8	27.8
Shared matting [14]	20.7	20.3	22.5	19.3

The best results are highlighted in bold.
KL, Kullback-Leibler; KNN, K-nearest neighbors; PDMS, pixel-level discrete multiobjective sampling; SDMS, sampling-based matting algorithm.

first and second in overall gradient error performance, respectively. The triple-objective sampling version of the pixel-level multiobjective global sampling image matting algorithm ranked first in the comparison of the average gradient error of trimaps with large unknown regions and manually labeled trimaps. The gradient error performance of the pixel-level multiobjective global sampling image matting algorithm was better in the triple-objective version than in the dual-objective version. This experimental result indicates that, on the one hand, the pixel-level multiobjective global sampling image matting algorithm can improve sampling performance by considering more sampling criteria. On the other hand, the algorithm can adaptively weigh multiple sampling criteria to obtain high-quality pixel samples. In addition, the gradient error of the superpixel-level multiobjective global sampling image matting algorithm was significantly higher than that of the pixel-level multiobjective global sampling image matting algorithm, which reflects that the superpixel sampling caused the problem of losing optimal foreground/background pixel pairs, resulting in a significant decrease in segmentation performance.

Fig. 1.5 shows a comparison of the alpha matte results obtained using different image matting algorithms. Fig. 1.5A is the input image; Fig. 1.5B is a local magnified image of the yellow box area in Fig. 1.5A; Fig. 1.5C is the local magnified image of the alpha matte obtained using the pixel-level multiobjective global sampling-based method; Fig. 1.5D is the local magnified image of the alpha matte obtained using the sparse coding-based method; Fig. 1.5E is the local magnified image of the alpha matte obtained using the KL-divergence-based sampling method; Fig. 1.5F is the local magnified image of the alpha matte obtained using the K–nearest neighbor-based segmentation method; Fig. 1.5G is the local magnified image of the alpha matte obtained using the comprehensive sampling-based method; Fig. 1.5H is the local magnified image of the alpha matte obtained using the color and texture weighted segmentation method; and Fig. 1.5I is the local magnified image of the alpha matte obtained using the global sampling-based method. The arrow in the figure indicates the region where the alpha matte error is large. The pixel-level multiobjective global sampling-based method obtains high-quality alpha matte with sharper edges compared with other image matting algorithms. This visual comparison result is consistent with the objective quantitative comparison results of the alpha matte. The pixel-level multiobjective global sampling-based method can obtain sharper alpha matte edges mainly because it uses the pixel-level sampling strategy. The edges of the foreground object often contain

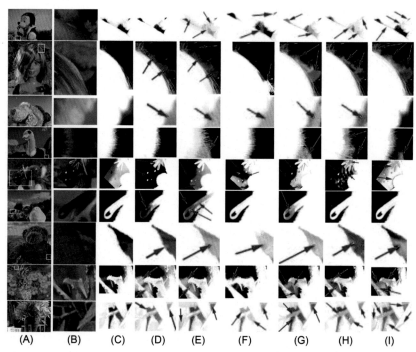

(A)   (B)   (C)   (D)   (E)   (F)   (G)   (H)   (I)

Figure 1.5 Comparison of extracted alpha mattes using different image matting algorithms. (A) Image. (B) Zoomed windows. (C) PDMS-based matting. (D) Graph-based sparse matting [8]. (E) KL divergence [7]. (F) KNN mattting [13]. (G) Comprehensive sampling [6]. (H) Weighted color and texture [5]. (I) Global sampling [3]. Arrows indicate regions of low-quality mattes. Failure cases are presented in the last two rows. *KL*, Kullback-Leibler; *KNN*, K-nearest neighbors; *PDMS*, pixel-level discrete multiobjective sampling.

shadows and highlights. The average color of the superpixel in the superpixel-level sampling cannot represent the color of the shadow and highlight regions since the shadow and highlight areas of known foreground/background pixels are very small. The pixel clustering process of the superpixel sampling often leads to the loss of the optimal foreground/background pixel pairs, resulting in large errors in the estimated alpha matte in the foreground edge region. The pixel-level multiobjective global sampling-based method abandons the pixel clustering operation, and each pixel can be collected as a sample. Multiobjective optimization can adaptively balance different sampling criteria. It can still collect high-quality pixel samples even when multiple sampling criteria conflict.

However, the pixel-level multiobjective global sampling-based image matting algorithm also has some limitations. As shown in Fig. 1.6,

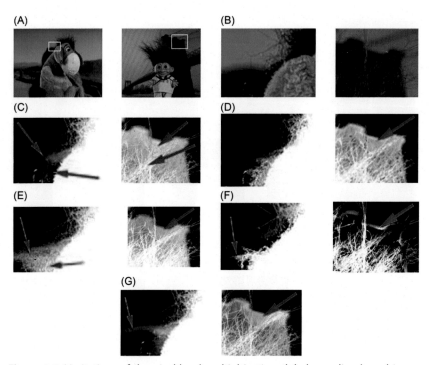

Figure 1.6 Limitations of the pixel-level multiobjective global sampling-based image matting algorithm. (A) Image. (B) Zoomed windows. (C) PDMS-based matting. (D) Graph-based sparse matting [8]. (E) KL divergence [7]. (F) KNN mattting [13]. (G) Comprehensive sampling [6]. (H) Weighted color and texture [5]. (I) Global sampling [3]. Arrows indicate regions of low-quality matters. Failure cases are presented in the last two rows. *KL*, Kullback-Leibler; *KNN*, K-nearest neighbors; *PDMS*, pixel-level discrete multiobjective sampling.

Fig. 1.6A is the input image; Fig. 1.6B is a locally enlarged image of the yellow box area in Fig. 1.6A; and Fig. 1.6C—G are matte masks locally enlarged image obtained by different matting algorithms. Fig. 1.6C is obtained by the pixel-level multiobjective global sampling-based image matting algorithm; Fig. 1.6D is obtained by the sparse coding-based matting algorithm; Fig. 1.6E is obtained by the KL-divergence sampling-based matting algorithm; Fig. 1.6F is obtained by the K-nearest neighbor matting algorithm; and Fig. 1.6G is obtained by the comprehensive sampling-based matting algorithm. In the case where the foreground and background color distributions overlap, the color of unknown pixels can be represented by both foreground and background pixel samples. The evaluation function used by the pixel-level multiobjective global sampling-based image matting algorithm cannot accurately select the optimal foreground/background pixel pairs in this situation, resulting in an error in the estimation of the matte mask.

1.1.3 Summary

This section introduces a multiobjective optimization sampling-based matting algorithm, focusing on the problems of slow speed in optimization-based free-sampling matting and low accuracy in sampling-based image matting algorithms. Unlike existing sampling strategies that assume that the optimal pixel pair may fall within a specific region, the introduced pixel-level multiobjective global sampling uses multiobjective optimization to provide a set of samples that can approximate the optimal solution for each unknown pixel pair, overcoming the problem of losing optimal pixel pairs due to unmet assumptions in existing algorithms. By organically combining optimization-based and sampling-based matting techniques, this method retains the high accuracy of optimization-based matting and the fast speed of sample-based matting. This section reveals conflicts in multiple sampling criteria and reasons for the loss of optimal pixel pairs due to the superpixel-level sampling. It also introduces the use of a parameter-free pixel-level multiobjective global sampling algorithm to solve these two problems. The multiobjective sampling strategy and the pixel-level sampling strategy are the core of this algorithm. The multiobjective sampling strategy addresses the conflict between multiple sampling criteria by modeling the multicriteria sampling problem as a discrete multiobjective optimization problem and using Pareto optimal solutions of the multiobjective optimization problem as pixel samples. By simultaneously minimizing color, spatial distance, and texture differences between known and unknown pixels, the multiobjective sampling strategy achieves adaptive balancing of multiple sampling criteria, which enables high-quality pixel samples to be collected even in the presence of sampling criteria conflicts. The multiobjective sampling strategy is easy to implement. It can be extended to more sampling criteria. The pixel-level sampling strategy avoids the incomplete sampling space by extending the sampling space to the set of all known foreground and background pixels, which allows each pixel in the known area to be collected as a pixel sample. Experimental results show that the candidate set of pixel pairs collected by the pixel-level multiobjective global sampling algorithm has a small cardinality and can achieve a small minimum alpha value absolute error. Based on the pixel-level multiobjective global sampling algorithm, this section introduces a matting algorithm based on pixel-level multiobjective global sampling. Objective and visual comparison experiments of matting alpha value errors show that this matting algorithm has significantly lower gradient errors in the foreground object edge area than existing matting algorithms and can obtain high-quality alpha mattes with sharp edges, while the computational time is comparable to that of existing sample-based image matting algorithms.

1.2 Strong collaboration of fuzzy logic and evolutionary computing

1.2.1 Overview of research progress

Image matting is an image and video processing technique that accurately extracts foregrounds by estimating their opacity. It has vast application potential in fields such as image composition, video post-production, and virtual reality. In image matting problems, pixel pairs directly determine the opacity of unknown pixels, so selecting appropriate pixel pairs is the core issue. The heuristic optimization-based image matting technology offers a feasible solution to the global search for pixel pair optimization problems. In this technology, the core process of selecting pixel pairs includes pixel pair evaluation and pixel pair optimization.

The goal of pixel pair evaluation is to accurately and quantitatively measure the quality of pixel pairs. High-quality pixel pairs correspond to unknown alpha values of pixels that are close to the corresponding values in the ground-truth alpha matte. The difficulty of pixel pair evaluation lies in the fact that the ground-truth alpha matte is unknown during the evaluation process, and that pixel pair evaluation needs to be robust to changes in the shape, color, lighting, and scene of the foreground object. Researchers have introduced various pixel pair evaluation criteria, such as low color distortion [2], spatial distance proximity [3], and texture similarity [5]. Recently, pixel pair evaluation methods have been used to improve the accuracy of evaluation by using multiple evaluation criteria. The evaluation functions of these evaluation methods often contain multiple evaluation items, each corresponding to an evaluation criterion. Multiple evaluation criteria items are constructed by linearly weighting [3,16] or simply nonlinearly combining multiple evaluation items (e.g., multiplying multiple evaluation items) [5−7] to construct the pixel pair evaluation function. An underlying assumption of the evaluation method is that multiple pixel pair evaluation criteria can be satisfied simultaneously. However, this assumption may not hold in complex situations. For example, in the case where the optimal foreground pixel is close to the unknown pixel and where the optimal background pixel is far from the unknown pixel, although the foreground pixel satisfies the proximity evaluation criterion, it is difficult for the optimal background pixel to satisfy this criterion.

Due to the uncertainty of the degree of satisfaction of the evaluation criteria, existing evaluation methods often cannot provide accurate evaluations in this situation. Current evaluation methods still cannot handle the uncertainty of the degree of satisfying multiple criteria. Pixel pair optimization is a challenge in pixel pair selection. On the one hand, the optimization objective function contains multiple evaluation criteria with uncertain satisfaction levels, making pixel pair optimization a complex combinatorial optimization problem. On the other hand, the large search space and the large number of decision variables involved in this problem make pixel pair optimization even more difficult.

Existing research has reduced the search space by sampling to achieve an approximate solution to the pixel pair optimization problem [2,3,6,7,14]. Although the sampling-based image cut algorithms have the advantage of low computational complexity, they suffer from the problem of inevitably losing the optimal samples due to changes in image cut scenes, lighting, and other factors. To solve this problem, recent research treats pixel pair optimization as a large-scale continuous optimization problem and implemented sampling-free pixel pair global search through large-scale heuristic optimization techniques. It theoretically avoids the problem of losing the optimal samples since this algorithm does not require sampling. Existing sampling-free image cut algorithms directly optimize the pixel pair evaluation function composed of multiple evaluation items. However, existing algorithms have not fully utilized the heuristic information provided by each evaluation item during the optimization process since the heuristic information of each evaluation item is hidden in the process of composing the objective function. The heuristic information provided by a single evaluation item can often guide heuristic optimization algorithms to escape local optima and approach global optima. An example is given in Fig. 1.7, where three evaluation items make up a complex multicriteria evaluation function, and the synthesized multicriteria evaluation function lacks sufficient heuristic information, making it difficult to escape local optima. The heuristic information provided by evaluation criterion 2 (as shown in Fig. 1.7) can guide heuristic optimization algorithms to approach the global optimum from the local optimum.

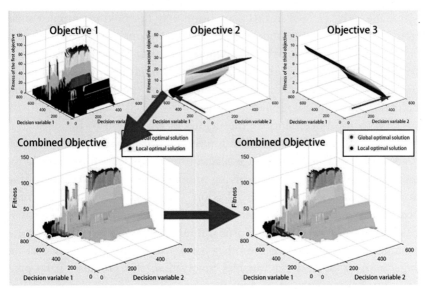

Figure 1.7 Example of a single evaluation item providing heuristic information to guide heuristic optimization algorithms to jump out of local optimal solutions and approach global optimal solutions in multicriteria evaluation functions. The red arrow in the figure indicates the search direction guided by heuristic information.

Currently, there are few studies on multiobjectivization of single-objective optimization problems. Some studies convert multicriteria single-objective optimization problems into multiobjective optimization problems. Knowles et al. [17] introduced two ways to multiobjectivize single-objective optimization problems: designing an additional optimization objective that has the same search space as the original problem or decomposing the original optimization problem into multiple sub-optimization problems by grouping the decision variables. Inspired by the work of Knowles et al., Greiner et al. [18] designed an auxiliary optimization objective for the framework structure optimization problem. Considering the uncertainty of high-quality pixels on features, it is difficult to design an auxiliary optimization objective that can provide rich heuristic information, and the approach of designing auxiliary optimization objectives is not suitable for pixel pair optimization problems with multiple evaluation criteria. Although the group-based multiobjectivization method can reduce the search space of the optimization problem, this method still does not utilize the heuristic information of a single evaluation criterion.

1.2.2 Scientific principles
1.2.2.1 Problem description
This section introduces the optimization problem of the pixel pair evaluation function involving multiple evaluation criteria and provides its mathematical model. Unlike the previously discussed model of large-scale pixel pair optimization problems, the evaluation function of the pixel pair optimization problem discussed in this section involves multiple evaluation criteria, including more than one evaluation item. To provide a concise expression without losing generality, the multicriterion pixel pair evaluation function is modeled as a combination of multiple evaluation items corresponding to each evaluation criterion. If the multicriterion evaluation function involves N_{obj} evaluation items, and $h^i_k(x)$ represents the ith evaluation item corresponding to the kth unknown pixel under the ith evaluation criterion, $i = 1, 2, \ldots, N_{obj}$, the multicriterion evaluation of the kth unknown pixel can be modeled as follows:

$$g_k(x_k) = H\left(h^1_k(x_k), h^2_k(x_k), \ldots, h^{N_{obj}}_k(x_k)\right) \tag{1.14}$$

where x_k represents the decision variable for the kth unknown pixel, and $H(h^1_k(x_k), h^2_k(x_k), \cdots, h^{N_{obj}}_k(x_k))$ represents the combination of N_{obj} evaluation criteria through linear or nonlinear methods. The optimization model for pixel pairs with multiple evaluation criteria is

$$G(X) = \sum_{k=1}^{N} H\left(h^1_k(x_k), h^2_k(x_k), \ldots, h^{N_{obj}}_k(x_k)\right) \tag{1.15}$$

where X represents the decision variables for the pixel pair optimization problem for the entire image, $X = (x_1, x_2, \ldots, x_N)^T$, and N represents the number of unknown pixels in the given segmentation problem.

1.2.2.2 Multiobjective collaborative optimization image matting algorithm based on fuzzy multicriteria evaluation and decomposition
In response to the uncertainty of the degree to which the pixel pair evaluation criteria are satisfied and the problem of low search accuracy faced by existing heuristically optimized image matting algorithms, this section introduces a multiobjective collaborative optimization image matting algorithm based on fuzzy multicriteria evaluation and decomposition. The algorithm includes two parts: fuzzy multicriteria evaluation method and decomposition-based multiobjective collaborative optimization algorithm. Considering that fuzzy mathematics can effectively deal with the

uncertainty of multiple-input and multiple-output systems [19], each evaluation criterion is modeled as a fuzzy membership function in the fuzzy multicriteria evaluation method. Fuzzy logic operations are utilized to synthesize the fuzzy multicriteria pixel pair evaluation function from multiple evaluation criteria. The introduced decomposition-based multiobjective collaborative optimization algorithm is based on the idea of the "variational method." The algorithm decomposes the pixel pair evaluation function of multiple criteria into multiple single-objective functions corresponding to each evaluation criterion and optimization objective. By modeling multiple objectives as a multiobjective optimization problem, multiple single-objective functions are optimized simultaneously using multiobjective optimization techniques, thereby fully utilizing the heuristic information of each evaluation criterion. Next, the introduced fuzzy multicriteria evaluation method is used to select the optimal pixel pair from the Pareto solution set. The fuzzy multicriteria pixel pair evaluation method and the decomposition-based multiobjective collaborative optimization algorithm will be introduced separately below.

1.2.2.2.1 Fuzzy multicriteria pixel pair evaluation method

In this section, a fuzzy multicriteria pixel pair evaluation method is introduced for the uncertainty of the degree of satisfaction of the pixel pair evaluation criteria. The fuzzy multicriteria evaluation of pixel pairs is achieved by evaluating the fuzzy affiliation degree of their belonging to the set of high-quality pixel pairs. The characteristics of high-quality pixel pairs are uncertain. Three widely used evaluation criteria are introduced to be modeled as three affiliation functions. The degree of affiliation of a pixel pair to the set of high-quality pixel pairs is measured by fuzzy logic operations.

The pixel pair evaluation criteria and the membership functions are introduced as follows. The first evaluation criterion is the reconstruction color error criterion. This criterion is used to evaluate whether a pixel pair can effectively reconstruct the color of an unknown pixel through a nonconvex combination of foreground and background colors. According to this criterion, the following color membership function is established by measuring the color distortion between the corresponding synthesized color of the pixel pair and the observed color:

$$h_k^1\left(x_k^{(F)}, x_k^{(B)}\right) = \exp\left(-\sigma_c \left\| C_k^{(U)} - \hat{\alpha} C_{x_k^{(F)}}^{(F)} - (1 - \hat{\alpha}) C_{x_k^{(B)}}^{(B)} \right\|\right) \qquad (1.16)$$

where $C_k^{(U)}$, $C_{x_k^{(F)}}^{(F)}$, and $C_{x_k^{(B)}}^{(B)}$ represent the RGB color vectors of the kth unknown pixel, foreground pixel, and background pixel in the pixel pair $(x_k^{(F)}, x_k^{(B)})$, respectively. σ_c is the color penalty coefficient, which is set to 0.11. A larger penalty coefficient indicates a stronger punishment for violating the evaluation criterion. Due to the mathematical model for image matting assuming that the color of all pixels in the image is a nonconvex combination of a foreground color and a background color, all unknown pixels should satisfy the criterion of color reconstruction error. The criterion of color reconstruction error is applicable to all unknown pixels.

The second and third criteria used are foreground and background pixel spatial distance criteria. The spatial distance criterion is based on the empirical assumption that the optimal pixel pairs are usually located in the region close to the unknown pixel. For these two evaluation criteria, we design the foreground pixel spatial position membership function and the background pixel spatial position membership function. The membership functions are given by Eqs. (1.17) and (1.18), respectively.

$$h_k^2\left(x_k^{(F)}, x_k^{(B)}\right) = \exp\left(-\sigma_s \left\| S_k^{(U)} - S_{x_k^{(F)}}^{(F)} \right\|^2\right) \qquad (1.17)$$

$$h_k^3\left(x_k^{(F)}, x_k^{(B)}\right) = \exp\left(-\sigma_s \left\| S_k^{(U)} - S_{x_k^{(F)}}^{(B)} \right\|^2\right) \qquad (1.18)$$

$S_k^{(U)}$ represents the kth unknown pixel's spatial coordinate vector. $S_{x_k^{(F)}}^{(F)}$ and $S_{x_k^{(B)}}^{(B)}$ denote the spatial coordinate vectors of the foreground and background pixels corresponding to the k unknown pixels in the $k \times k$ pixel pairs, respectively. σ_s is the spatial position penalty coefficient with a value of 0.17. The pixel space distance criterion is applicable to the matting task where there are no holes inside the foreground object. When there are holes inside the foreground object, the foreground and background pixel space distance criteria may not be satisfied at the same time. Fig. 1.8 shows an example where the red and blue solid lines represent the edges of the known foreground and known background regions in the trimap, respectively, and the yellow asterisks represent the unknown pixels. The optimal foreground and background pixels that correspond to the unknown pixel are marked with red and blue asterisks, respectively. In this example, because there are holes inside the foreground object, the optimal background pixel's spatial distance to the unknown pixel is far, satisfying the foreground pixel space distance criterion but not the background pixel

Figure 1.8 Example where the foreground and background pixel spatial distance criteria cannot be simultaneously satisfied.

space distance criterion. Similarly, an example can be given where the optimal pixel pair satisfies the background pixel space distance criterion but not the foreground pixel space distance criterion.

The fuzzy multiple criteria pixel pair evaluation method uses fuzzy aggregation operations [20] to merge the values of multiple membership functions into adapted values for pixel pairs. High-quality pixel pairs may not meet both spatial criteria at the same high level (i.e., in a particular case, one pixel may be far from an unknown pixel in space). To deal with the uncertainty of high-quality pixel pairs meeting spatial criteria, the fuzzy multiple criteria evaluation method adopts an average aggregation operation to merge the membership values of the two spatial criteria, reducing the evaluation error of pixel pairs when one spatial position criterion is low. If a and b are two membership values, the average aggregation operation can be expressed as follows:

$$f_1(a, b) = 0.5 \cdot (a + b) \qquad (1.19)$$

The aggregated fuzzy spatial membership will be merged with the color membership to generate the final adaptation value for each pixel. Considering that the Einstein product operation [2][5] can approximate the algebraic product of fuzzy subsets well (see footnote 5), the Einstein

---

[5] Let $A = \int \mu_A(x)/x, B = \int \mu_B(y)/y$ be two fuzzy subsets of a corresponding determined set Z, where $\mu_A(x)$ and $\mu_B(y)$ are fuzzy membership functions, and $*$ represents a binary operation on the determined set Z. Then $*$ can be extended to algebraic product operation on the fuzzy sets A and B [22] $A * B = \left(\int \mu_A(x)/x\right) * \left(\int \mu_B(y)/y\right) = \int (\mu_A(x) \wedge \mu_B(y))/(x * y)$, where $\wedge$ represents the minimum value operation.

product operation will be used to further aggregate the color membership values with the aggregated fuzzy spatial membership. The definition of the Einstein product operation is given by the following equation:

$$f_2(a,b) = \frac{a \cdot b}{1 - (1-a) \cdot (1-b)} \quad (1.20)$$

In summary, the fuzzy multicriteria pixel pair evaluation function can be expressed as follows:

$$g_k(i,j) = f_2\left(h_k^1(i,j), f_1\left(h_k^2(i,j), h_k^3(i,j)\right)\right) \quad (1.21)$$

Combining Eqs. (1.19)−(1.21), we get

$$g_k(i,j) = \frac{h_k^1(i,j) \cdot 0.5\left(h_k^2(i,j) + h_k^3(i,j)\right)}{1 - \left(1 - h_k^1(i,j)\right) \cdot \left(1 - 0.5 \cdot \left(h_k^2(i,j) + h_k^3(i,j)\right)\right)} \quad (1.22)$$

1.2.2.2.2 Multiobjective collaborative optimization algorithm based on decomposition

To fully utilize the heuristic information provided by each criterion in the multicriterion pixel pair evaluation function, this section introduces a multiobjective collaborative optimization algorithm based on decomposition [21]. Existing free-sampling image matting algorithms based on heuristic optimization use large-scale heuristic optimization algorithms oriented towards single-objective optimization to directly optimize multicriterion objective functions. However, as shown in Fig. 1.7, the heuristic information generated during the optimization process of individual evaluation criteria is lost in the process of aggregating the results of multiple evaluation criteria to produce fitness values. This loss of information would guide the population to jump out of the local optimums and approach the global optimal solutions during the iteration process. The basic idea of the introduced multiobjective collaborative optimization algorithm based on decomposition is to convert the multicriterion pixel pair optimization problem into a multiobjective problem, with each criterion corresponding to one optimization objective. This decomposition strategy can effectively utilize the heuristic information generated by each criterion and utilize existing mature multiobjective optimization techniques. Fig. 1.9 shows the basic process of the multiobjective collaborative optimization algorithm based on decomposition.

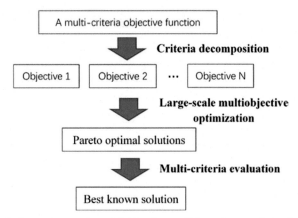

Figure 1.9 Basic process of multiobjective collaborative optimization algorithm based on decomposition.

The introduced multiobjective collaborative optimization algorithm based on decomposition mainly consists of three steps:

1) Use the introduced multicriteria evaluation function decomposition optimization strategy to decompose the multicriteria single-objective evaluation function into multiple single-criteria single-objective functions based on the average criterion.

2) Use multiobjective heuristic optimization algorithms to optimize multiple single-criteria single-objective functions simultaneously. In the optimization process, the problem is decomposed into multiple subproblems by grouping the decision variables into neighborhoods, and collaborative optimization is achieved based on the correlation between the subproblems. Based on the divide-and-conquer principle, multiobjective heuristic optimization algorithms divide the decision variables into groups based on their unknown pixel space correlation and optimize the solution through competitive optimization of the optimal pixels within the group, thus efficiently solving the pixel pairing optimization problem.

3) Use fuzzy multicriteria evaluation method to evaluate each pixel pair in the Pareto optimal solution set obtained in Step 2 of the multiobjective optimization problem, then select the optimal pixel pair.

Algorithm 1.2 provides a specific implementation method for the multiobjective collaborative optimization algorithm based on decomposition. Γ is the set of optimized objectives obtained through multicriteria

---

ALGORITHM 1.2 The multiobjective collaborative optimization algorithm based on decomposition.

Input a large-scale multicriteria function

$$G(X) = \sum_{K=1}^{C} H(h_k^1(X), h_k^2(X), \cdots, h_k^{N_{obj}}(x_k))$$

Output X_{best}

    // Criteria decomposition:

1: $\Gamma \leftarrow \varnothing$

2: **for** $i = 1$ to N_{obj} **do**

3:    $\Gamma \leftarrow \Gamma \cup h_k^i(X)$

4: **end for**

    // multi-objective optimization:

5: $\{\Lambda_1, \Lambda_2, \ldots, \Lambda_N\} \leftarrow O(\Gamma, MOO, 1)$

    // Multi-criteria evaluation:

6: **for** $i = 1$ to N **do**

7:    $x_{best}^i \leftarrow rand()$

8:    **for** $m = 1$ to $|\Lambda_i|$ **do**

9:      **if** $g_i(x_m^i) < g_i(x_{best}^i)$ **then**

10:     $x_{best}^i \leftarrow x_m^i$

11:     **end if**

12:    **end for**

13: **end for**

14: $X_{best} \leftarrow (X_{best}^1, X_{best}^2, \ldots, X_{best}^N)$

15: **return** X_{best}

---

decomposition. Λ_i represents the Pareto optimal solution of the ith subproblem obtained. x_{best}^i represents the current optimal solution of the ith subproblem during the algorithm iteration process. x_m^i represents the mth solution in the Pareto optimal solution of the ith subproblem. O represents the collaborative multiobjective optimization algorithm based on neighborhood grouping. MOO represents the multiobjective optimization algorithm. $rand()$ represents a random real number generator, with the generated random number range being $[0,1]$.

The following introduces the multicriteria evaluation function decomposition optimization strategy and the neighborhood grouping collaborative multiobjective optimization strategy involved in the decomposition-based multiobjective collaborative optimization algorithm.

Different from existing works, the introduced algorithm of decomposition-based multiobjective collaborative optimization converts the complex multicriterion single-objective optimization problem into a multiobjective optimization problem through the strategy of decomposing the fuzzy multicriterion pixel pair evaluation function into an optimization target corresponding to each criterion. All optimization targets corresponding to each criterion are modeled as a multiobjective optimization problem. Each evaluation item in the fuzzy multicriterion pixel pair evaluation is modeled as an optimization objective positively correlated with the fitness value, and the evaluation item negatively correlated with the fitness value can be transformed into a positively correlated optimization objective by multiplying it by -1. A single optimization objective can provide rich heuristic information in the optimization process. This heuristic information guides the heuristic optimization algorithm to approach the optimal solution. Through multiobjective optimization for each target corresponding to the criterion, the performance of the heuristic optimization algorithm can be improved by fully utilizing the heuristic information provided by each criterion.

The strategy of neighborhood grouping collaborative multiobjective optimization is adopted to solve large-scale multiobjective optimization problems generated by the decomposition of multicriteria evaluation functions, based on decomposition-based multiobjective collaborative optimization algorithm. Large-scale optimization problems usually face the problem of dimensionality catastrophe—as the dimensionality of decision variables increases, the search space increases exponentially. The strategy of grouping decision variables has been proven to be effective in solving large-scale optimization problems. To improve the optimization efficiency of pixel pairs, this section introduces a neighborhood grouping collaborative multiobjective optimization strategy based on the local smoothness of images. The strategy includes two steps: neighborhood grouping and multiobjective collaborative optimization. Fig. 1.10 shows the framework of the neighborhood grouping collaborative multiobjective optimization strategy.

We noticed that the optimal pixel pairs corresponding to unknown pixels within a local area have little variation [13,22,23]. We call this the local smoothness property of the image. The introduced grouping strategy is based on this property and achieves efficient grouping of decision variables based on strong spatial correlation. Specifically, for a given unknown pixel, the decision variables for pixel pairs corresponding to unknown pixels in a 9×9 neighborhood centered at that point are grouped together. Unlike existing grouping strategies for black-box optimization, this grouping uses prior knowledge of natural image matting, the local

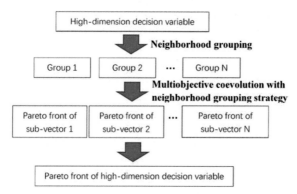

Figure 1.10 Basic process of decomposition-based multiobjective collaborative optimization algorithm.

smoothness property, to achieve fast and accurate grouping of decision variables. Grouping strategies for black box optimization can consume a large number of computation resources. Compared to existing general grouping methods, the introduced grouping strategy saves limited computational resources.

In the process of multiobjective collaborative optimization, a set of unknown pixel pairs is selected at equal intervals for independent optimization in each group. The sampling interval of unknown pixels is set to one pixel. A multiobjective optimization algorithm is used to independently optimize the selected unknown pixel pairs. To use existing multiobjective optimization algorithms with strong optimization capabilities, such as nondominated sorting genetic algorithm [24] and decomposition-based multiobjective evolutionary algorithm [25], the pixel pair combination optimization problem is relaxed to a continuous problem. That is, the decision space is mapped from $\{1,2,\ldots,|\Omega_F|\}^c \times \{1,2,\ldots,|\Omega_B|\}^c$ to $[1,|\Omega_F|]^c \times [1,|\Omega_B|]^c$. The obtained solutions in the continuous space are mapped to the discrete space through rounding to evaluate the pixel pairs. Decomposing the multi-criteria single-objective optimization problem into multiple single-objective optimization problems is to fully utilize the heuristic information provided by each evaluation criterion and improve the optimization ability of heuristic optimization algorithms. The decomposition-based multiobjective collaborative optimization algorithm independently solves the selected multiobjective optimization problems using the decomposition-based multiobjective evolutionary algorithm introduced by Zhang et al. [25]. The decomposition-based multiobjective evolutionary algorithm decomposes the multiobjective optimization problem into multiple single-objective optimization subproblems by weighting multiple objectives and optimizes the subproblems simultaneously [25]. The mechanism of the decomposition-based multiobjective evolutionary algorithm determines that some subproblems are the same as the single-objective optimization problems corresponding to a single criterion, while some subproblems may be highly similar to the multicriteria optimization problem. The former can fully utilize the heuristic information provided by a single criterion, and the latter ensures that the optimization goal of the population is consistent with the multicriteria evaluation goal of the pixel pairs. After obtaining the Pareto optimal solution for each subproblem in the group, the Pareto optimal solution set with the best result in the group is generated by further "competing" the corresponding Pareto optimal solutions of all subproblems in the group for any unknown pixel and removing the dominated solutions. The specific implementation of the neighborhood group collaborative multiobjective optimization algorithm is shown in Algorithm 1.3. $\mathcal{N}(S_i^{(U)}, S_j^{(U)})$ is the indicator function of the spatial neighborhood. The function value is true when $S_i^{(U)}$ is the $S_j^{(U)}$ neighborhood, and false otherwise. Sub_j represents the pixel pair optimization subproblem of unknown region pixel j.

Under the local smoothness hypothesis, the solution to the optimization problem of multiple pixel pairs within a group (i.e., a local area) can be approximated by solving the optimization problem of a single pixel pair. However, considering the instability of random algorithms and the complexity of the pixel pair optimization problem, the collaborative multiobjective optimization algorithm improves the stability of generating alpha mattes by optimizing redundant pixel pairs within a group. The competition among different unknown pixels optimized in the neighborhood improves the quality of optimization and makes the obtained alpha matte smoother.

The computational complexity of the collaborative multiobjective optimization algorithm based on neighborhood grouping is analyzed below. In the analysis process, the optimization of the subproblem is taken as the basic operation. That is, the computational complexity of optimizing a subproblem is $O(1)$. As only unknown pixels whose coordinates can be divided by ρ are selected for subproblem optimization, assuming that the unknown pixels are uniformly distributed in the trimap, the probability of a subproblem being selected for optimization is $\frac{\lfloor h/\rho \rfloor}{h} \cdot \frac{\lfloor w/\rho \rfloor}{w}$, where h and w represent the length and width of the input image, respectively.

ALGORITHM 1.3 The collaborative multiobjective optimization algorithm based on neighborhood grouping.

Input: *MOP, MOO,* and ρ.

Output: $\{\Lambda_1,\Lambda_2,\ldots,\Lambda_N\}$

1: **for** $i = 1$ to N **do**

2:     //*neighborhood grouping*

3:     $\Omega_N \leftarrow \varnothing$

4:     **for** $j = 1$ to N **do**

5:         **if** $N\left(S_i^{(U)}, S_j^{(U)}\right)$ **then**

6:             $\Omega_N \leftarrow \Omega_N \cup j$

7:         **end if**

8:     **end for**

9:     //*Optimizing selected subproblems in the group:*

10:     $\Lambda_i \leftarrow \varnothing$

11:     **for** each $j \in \Omega_N$ **do**

12:         **if** the coordinate of pixel j x,y are dividable by ρ **then**

13:             **if** $\Lambda_j = \varnothing$ **then**

14:                 $\Lambda_j \leftarrow MOO(Sub_j)$

15:             **end if**

16:         **end if**

17:     **end for**

18: **end for**

19: //*Competing in the group*

20: **for** $i = 1$ to N **do**

21:     **if** $\Lambda_i \neq \varnothing$ **then**

22:         **for** each $j \in \Omega_N$ **do**

23:             **if** $\Lambda_j = \varnothing$ **then**

24:                 $\Lambda_j \leftarrow \Lambda_i$

25:             **else**

26:                 $\Lambda_j \leftarrow \Lambda_j \cup$ the Pareto solutions of Λ_i

27:             **end if**

28:         **end for**

29:     **end if**

30: **end for**

Since $\frac{\lfloor h/\rho \rfloor}{h} \cdot \frac{\lfloor w/\rho \rfloor}{w} \leq \frac{h/\rho}{h} \cdot \frac{w/\rho}{w} = 1/\rho^2$, the expected number of selected subproblems is no more than N/ρ^2. Each selected subproblem is optimized only once, so the average time complexity of the algorithm is $O(N/\rho^2)$, where ρ is not less than 1. Therefore, the time complexity can be represented as $O(N)$.

1.2.2.3 Experimental results and discussion

Experiments were conducted to select a suitable multiobjective optimization algorithm based on fuzzy multicriteria evaluation and decomposition in the multiobjective collaborative optimization image matting algorithm. Then the effectiveness of the introduced algorithm is verified through four experiments. The first experiment demonstrates the performance stability of the multiobjective collaborative optimization image matting algorithm using different multiobjective optimization algorithms. The second experiment is used to verify whether the fuzzy multicriteria evaluation for pixel pairs can effectively handle uncertainty in multicriteria evaluation. In the third experiment, the decomposition-based multiobjective collaborative optimization algorithm was compared with other advanced large-scale optimization algorithms to verify its effectiveness. The fourth experiment verifies whether fuzzy multicriteria evaluation and decomposition-based multiobjective collaborative optimization image matting algorithm can provide high-quality alpha mattes.

In the experiments, the images and trimaps provided by Rhemann et al.'s benchmark dataset [10] were employed. This benchmark dataset provides 35 color images, each of which contains two machine-generated trimaps corresponding to large and small unknown regions. Among them, 27 images provide ground-truth alpha mattes. The ground-truth alpha mattes for the remaining 8 images are not publicly available and are used for online image matting algorithm performance evaluation. Additional manually labeled trimaps are provided for these eight images.

The color penalty factor and spatial penalty factor in the fuzzy multicriteria pixel pair evaluation are set to 0.11 and 0.17, respectively. The unknown pixel sampling interval ρ is set to 1. All experiments were run on a server with an Intel Xeon E5 2.4 GHz processor and 32 GB of memory. All algorithms involved in the experiments were implemented in MATLAB.

1.2.2.3.1 Multiobjective optimization algorithm selection experiment

Multiobjective collaborative optimization algorithm based on fuzzy multicriterion evaluation and decomposition for image matting transforms multiple criterion single-objective optimization problems into multiple single-objective optimization problems and optimizes them simultaneously using a multiobjective optimization algorithm. Considering that many strong multiobjective optimization algorithms are available, this algorithm uses them for optimization. This experiment was conducted to evaluate the effects of different multiobjective optimization algorithms on the performance of the introduced algorithm, thereby verifying its effectiveness.

This experiment selected three widely used multiobjective optimization algorithms: the multiobjective evolutionary algorithm based on decomposition (MOEA/D) introduced by Zhang et al. [25], the multiple objective particle swarm optimization (MOPSO) introduced by Coello et al. [26], and the fast elitist nondominated sorting genetic algorithm (NSGA-II) introduced by Deb et al. [24]. The parameters involved in the three algorithms were set according to the corresponding literature recommendations. To comprehensively demonstrate the impact of different algorithms on the performance of the fuzzy multicriteria evaluation and decomposition-based multiobjective cooperative optimization image matting algorithm, 27 images with ground-truth alpha mattes and two types of trimaps with large and small unknown regions were used as experimental data. The experiment used the mean square error to quantitatively evaluate the obtained alpha matte.

Tables 1.5 and 1.6 summarize the mean squared error of alpha mattes using a fuzzy multicriteria evaluation and decomposition-based multiobjective collaborative optimization algorithm with different

Table 1.5 Mean square error of the alpha matte obtained by different multiobjective optimization algorithms used in the fuzzy multicriteria evaluation and decomposition-based multiobjective collaborative optimization image matting algorithm, which was obtained using a trimap with a large unknown area.

Image name	GT01	GT02	GT03	GT04	GT05	GT06	GT07	GT08	GT09
MOEA/D	58.8	143.1	206.5	673.5	**89.2**	**141.1**	**54.7**	774.2	**116.9**
MOPSO	**56.5**	160.7	**178.8**	654.3	115.7	217.3	68.1	**757.5**	127.0
NSGA-II	57.4	**139.7**	191.0	**637.1**	107.9	157.6	60.1	759.4	122.3
Image name	**GT10**	**GT11**	**GT12**	**GT13**	**GT14**	**GT15**	**GT16**	**GT17**	**GT18**
MOEA/D	**240.0**	381.3	99.9	463.1	**114.1**	299.3	2282.3	90.2	89.9
MOPSO	266.8	377.4	108.7	453.7	143.6	320.7	2734.1	109.5	121.2
NSGA-II	244.5	**373.5**	**96.2**	**449.5**	117.8	**280.9**	2525.3	95.7	101.0
Image name	**GT19**	**GT20**	**GT21**	**GT22**	**GT23**	**GT24**	**GT25**	**GT26**	**GT27**
MOEA/D	**199.6**	110.8	826.4	**74.9**	**87.7**	648.6	1786.2	1689.5	2797.3
MOPSO	299.6	121.4	**790.5**	87.8	110.6	615.2	1844.6	1598.8	**2273.5**
NSGA-II	240.6	111.6	813.6	83.4	97.3	**583.2**	1826.3	**1589.8**	2396.9

Note: 1. MOEA/D represents a decomposition-based multiobjective evolutionary algorithm [25].
2. MOPSO stands for multiobjective particle swarm optimization algorithm [26].
3. NSGA-II stands for nondominated sorting genetic algorithm II, a fast and elitist nondominated sorting genetic algorithm [24].
4. The best results are highlighted in bold.

Table 1.6 Mean square error of the alpha matte obtained by different multiobjective optimization algorithms used in the fuzzy multicriteria evaluation and decomposition-based multiobjective collaborative optimization image matting algorithm based on the small unknown area trimap.

Image name	GT01	GT02	GT03	GT04	GT05	GT06	GT07	GT08	GT09
MOEA/D	**29.4**	**67.3**	**125.1**	**390.6**	**36.9**	**71.8**	**37.1**	613.8	**99.3**
MOPSO	31.8	80.7	142.3	430.5	62.4	105.2	47.7	605.2	106.2
NSGA-II	31.5	69.7	126.8	400.6	43.8	83.3	42.2	**585.3**	99.6

Image name	GT10	GT11	GT12	GT13	GT14	GT15	GT16	GT17	GT18
MOEA/D	149.7	213.3	59.3	**228.3**	**63.1**	181.2	927.0	**58.6**	**44.0**
MOPSO	165.8	234.3	63.1	272.9	87.3	181.4	1325.5	76.2	74.5
NSGA-II	**145.4**	**198.2**	**53.2**	235.8	66.1	**165.4**	**815.6**	65.6	50.6

Image name	GT19	GT20	GT21	GT22	GT23	GT24	GT25	GT26	GT27
MOEA/D	**76.3**	50.4	**349.6**	40.8	50.7	359.7	**1245.8**	1003.0	1311.0
MOPSO	122.3	58.0	434.6	53.8	66.0	363.6	1302.6	1053.4	1136.6
NSGA-II	94.1	**49.4**	363.7	44.0	56.0	**328.5**	1270.6	**990.6**	1087.8

Note: 1. MOEA/D represents a decomposition-based multiobjective evolutionary algorithm [25].
2. MOPSO stands for multiobjective particle swarm optimization algorithm [26].
3. NSGA-II stands for nondominated sorting genetic algorithm II, a fast and elitist nondominated sorting algorithm [24].
4. The best results are highlighted in bold.

multiobjective optimization algorithms on trimap with larger and smaller unknown areas, respectively. In the case of a larger unknown area, the version using a decomposition-based multiobjective evolutionary algorithm performed better. Of the 27 images, 17 in the dataset perform better with the multiobjective particle swarm optimization algorithm, and 14 perform better with the fast elite nondominated sorting genetic algorithm. In the smaller trimap with unknown regions, the multiobjective evolutionary algorithm based on decomposition performs better in 25 of the 27 images compared to the multiobjective particle swarm optimization algorithm, and the multiobjective evolutionary algorithm based on decomposition performs better in 17 of the 27 images compared to the fast elite nondominated sorting genetic algorithm. Given that the fuzzy multicriteria evaluation and decomposition-based multiobjective collaborative optimization algorithm achieved lower mean square errors on most images when using the decomposition-based multiobjective evolutionary algorithm, the algorithm chooses this as its required multiobjective optimization operator and sets its parameters according to the recommendations in reference [25].

1.2.2.3.2 Fuzzy multicriteria pixel pair evaluation accuracy experiment

The design objective of this experiment is to test whether fuzzy multicriteria pixel-pair evaluation can accurately evaluate pixel pairs in situations where the degree of satisfaction for multiple evaluation criteria is uncertain. In this experiment, fuzzy multicriteria pixel-pair evaluation was compared with a popular evaluation method [16] that uses the same evaluation criteria and evaluated their performance on uncertain samples and multiple images. The experimental data consist of 27 images from the benchmark dataset with corresponding large unknown area trimaps and ground-truth alpha mattes.

This experiment was conducted to assess the accuracy of different evaluation methods in selecting pixel pairs for observing and evaluating the accuracy of the alpha matte. The experimental results were obtained through three steps: generating candidate pixel pairs using the global sampling algorithm [3], evaluating pixel pairs using different evaluation methods, and estimating the alpha matte using the pixel pairs with the best fitness value.

Firstly, we explored the efficacy of the multiobjective collaborative optimization algorithm on uncertain samples. An image called GT04 was selected as an example with multiple evaluation criteria with uncertain degrees, encompassing 109,718 unknown pixels. Two evaluation methods were used to select the optimal pixel pair to obtain the alpha matte, as shown in Fig. 1.11. Fig. 1.11A is the input image, where the boundaries of known foreground and known background regions are marked in red and blue, respectively. The locally magnified area is marked in yellow. Fig. 1.11B is the ground-truth alpha mattes, Fig. 1.11C is the alpha matte obtained using the fuzzy multicriteria pixel pair evaluation method, and Fig. 1.11D is the alpha matte obtained using a popular evaluation method with the same evaluation criteria as the introduced method. It can be seen from the figure that the alpha matte estimated by the evaluation method in literature [16] has a large error compared to the ground-truth alpha mattes in the flag area, while the error of the alpha matte estimated by the fuzzy multicriteria pixel pair evaluation method is relatively small. As shown in Fig. 1.11A, the unknown pixel area on the left of the blue flag is far from the optimal foreground pixel, so the spatial proximity criterion has a low degree of satisfaction. The fuzzy multicriteria pixel pair evaluation method explores the uncertainty of multiple evaluation criteria and improves the accuracy of evaluation, obtaining an accurate alpha matte.

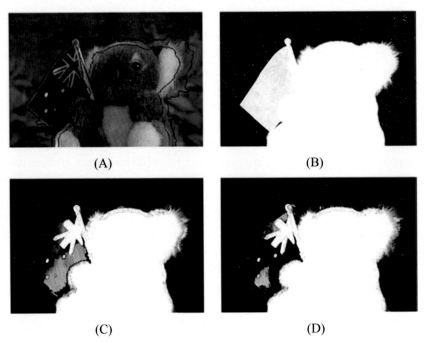

(A)   (B)

(C)   (D)

Figure 1.11 Alpha mattes obtained by using different pixel pair evaluation methods. (A) Input image. The boundaries of known foreground and known background regions are marked in red and blue, respectively. (B) Ground-truth alpha matte. (C) The alpha matte obtained by the introduced fuzzy multicriteria pixel pair evaluation method. (D) A popular pixel pair evaluation method [16] using the same evaluation criterion as the introduced methods.

The accuracy of the fuzzy multicriteria pixel pair evaluation method on average will be further discussed through statistical analysis. In this experiment, the evaluation performance of the introduced method and existing methods is quantitatively compared by the average mean squared error of the alpha value corresponding to the different pixel pairs selected by different evaluation methods. The calculation method of average mean squared error is as follows: Firstly, the pixel pair evaluation method is employed to select the best pixel pair from 2,893,894 unknown pixels in 27 images, and the mean squared error between the alpha value of selected pixel pairs and the corresponding value in the ground–truth alpha mattes is calculated. Then the average mean squared error of all unknown pixels is calculated. The corresponding average mean squared error of the fuzzy multicriteria pixel pair evaluation method is 603.99, which is lower than 622.77 obtained by the extant pixel pair evaluation method. The small average mean squared error indicates that the fuzzy multicriteria pixel pair evaluation method can correctly evaluate pixel pairs in various situations. In summary, this evaluation method effectively improves the accuracy of pixel pair evaluation in uncertain samples and on average. The experimental results show that the fuzzy membership function and fuzzy aggregation operator used in the fuzzy multicriteria pixel pair evaluation method effectively deal with the uncertainty of multiple evaluation criteria satisfaction.

1.2.2.3.3 Comparative experiment on optimization performance of multiobjective collaborative optimization algorithm based on decomposition

The main purpose of this experiment is to verify whether the decomposition-based multiobjective collaborative optimization algorithm can use heuristic information corresponding to each evaluation criterion to improve the quality of pixel pair optimization under limited computational resources. The experimental data include 27 images from the benchmark dataset that corresponds to ground-truth alpha mattes and their corresponding trimaps.

In the experiment, three advanced large-scale heuristic optimization algorithms were used as benchmarks for pixel pair optimization: differential grouping 2 (DG2) [27], competitive swarm optimization (CSO) [28], and cooperatively coevolving particle swarm optimization (CCPSO) [29]. In addition, to verify the effectiveness of the multicriteria evaluation function decomposition optimization strategy and the neighborhood grouping-based multiobjective optimization strategy, the competitive swarm optimization algorithm was combined with the aforementioned strategies. Fuzzy multicriteria evaluation and decomposition-based multiobjective collaborative optimization image matting algorithm and the four heuristic optimization algorithms were used to solve the large-scale pixel pair optimization problem. The maximum number of evaluations during the solution process was limited to 5000, and the mean squared error between the pixel pair corresponding to the alpha matte computed by different methods and the ground-truth alpha mattes was observed. The smaller the mean squared error of the alpha matte, the better the optimization performance of the heuristic optimization algorithm. Table 1.7 shows the mean squared errors of the alpha mattes obtained by the multiobjective collaborative optimization image matting algorithm based on fuzzy multicriteria evaluation and decomposition the

Table 1.7 Mean square error of alpha mattes obtained by a multiobjective collaborative optimization image matting algorithm based on fuzzy multicriteria evaluation and decomposition with three heuristic optimization algorithms on 27 images.

Image name	GT01	GT02	GT03	GT04	GT05	GT06	GT07	GT08	GT09
Multiobjective collaborative optimization image matting algorithm	58.8	**143.1**	**206.5**	**673.5**	**89.2**	**141.1**	**54.7**	**774.2**	**116.9**
CSO-NG	**54.5**	157.9	217.2	677.4	99.4	201.7	64.9	833.9	135.7
CSO [28]	731.9	3645.5	1048.2	3261.9	1136.9	2420.3	1037.1	2857.7	2475.3
CCPSO [29]	734.4	3739.6	1043.2	3204.5	1232.4	2501.1	1105.3	2861.2	2432.7

Image name	GT10	GT11	GT12	GT13	GT14	GT15	GT16	GT17	GT18
Multiobjective collaborative optimization image matting algorithm	240	**381.3**	**99.9**	463.1	**114.1**	**299.3**	**2282.3**	**90.2**	**89.9**
CSO-NG	254.5	394.3	114.3	**447.3**	169.6	337.5	2739.9	108.3	149.2
CSO [28]	2530.5	3905.1	342.2	5771.3	1104.4	1728.8	5525.7	1365.5	2451.8
CCPSO [29]	2609.6	3886.7	386.8	5859	1485.9	1821.9	5442.8	1695.9	2448.2

Image name	GT19	GT20	GT21	GT22	GT23	GT24	GT25	GT26	GT27
Multiobjective collaborative optimization image matting algorithm	**199.6**	110.8	826.4	**74.9**	**87.7**	648.6	**1786.2**	1689.5	2797.3
CSO-NG	256	**109.5**	**763.9**	82.1	100.4	**620.9**	1850.6	**1647.2**	**2593.2**
CSO [28]	1603.9	757.5	4475.6	1353	1559.3	3817.6	4129	6741.4	7202.3
CCPSO [29]	1547	779.2	4486.4	1353.7	1804.1	3785.9	3878.7	7362.1	7302.5

Note: 1. CSO-NG stands for competitive swarm optimization algorithm based on neighborhood grouping synergy.
2. CSO stands for competitive population optimization algorithm [28].
3. CCPSO stands for co-evolutionary particle swarm optimization algorithm [29].
4. The best results are highlighted in bold.

three heuristic optimization algorithms on the 27 images. It is worth noting that the DG2 algorithm ran out of 32 GB of memory during the optimization process and could not be used in this experiment.

The average mean square error of the introduced multiobjective collaborative optimization algorithm based on fuzzy multicriteria evaluation and decomposition for image matting is 538.5. The competitive group optimization algorithm based on neighborhood grouping collaboration has an average mean square error of 562.3. The competitive group optimization algorithm scores 2777.0 on average. The collaborative evolution particle swarm optimization algorithm yields an average mean square error of 2844.1. The introduced algorithm achieved lower alpha matte mean square error than existing large-scale heuristic optimization algorithms on all 27 images. This experimental result shows that multiobjective collaborative optimization algorithm based on fuzzy multicriteria evaluation and decomposition has higher optimization accuracy for pixel pair optimization problems compared to existing large-scale heuristic optimization algorithms. In addition, the multiobjective collaborative optimization algorithm based on fuzzy multicriteria evaluation and decomposition achieved a lower mean square error on 20 out of 27 images, as well as a lower average mean square error compared to the competitive group optimization algorithm. This competitive group optimization algorithm applies the multiobjective optimization strategy based on neighborhood group collaboration. Since both algorithms use the multiobjective optimization strategy based on neighborhood grouping collaboration for subproblem division and collaborative optimization, the difference in performance between the two originates from the optimization operator. This experimental result demonstrates that the multicriteria evaluation function decomposition optimization strategy fully utilizes the heuristic information provided by each evaluation criterion, thereby improving the diversity of the population. The diversity of the population can prevent heuristic optimization algorithms from getting stuck in local optimal solutions in pixel pair optimization problems with many local optimal solutions, and guide them to approach global optimal solutions. Therefore, multiobjective collaborative optimization image matting algorithm achieves better optimization quality than existing advanced large-scale optimization algorithms.

It is worth mentioning that the alpha matte mean square error obtained by the two image matting algorithms using the multiobjective optimization strategy based on neighborhood grouping collaboration (i.e., multiobjective collaborative optimization image matting algorithm and the competitive group optimization algorithm based on neighborhood grouping collaboration) is significantly lower than other large-scale heuristic optimization algorithms. The significant performance difference between algorithms using and not using the multiobjective optimization strategy based on neighborhood grouping collaboration reveals that the pixel pair optimization problem is a decomposable problem. It also verifies that the multiobjective optimization strategy based on neighborhood grouping collaboration can perform fast and effective grouping of decision variables in pixel pair optimization problems. The significant performance improvement brought by the multiobjective optimization strategy based on neighborhood grouping collaboration can be attributed to the prior knowledge of the problem: the local smoothness property decomposes the complex large-scale pixel pair optimization problem into multiple relatively simple subproblems.

1.2.2.3.4 Multiobjective collaborative optimization method for image matting based on fuzzy multicriteria evaluation and decomposition: comparative experiment of image matting performance

This experiment aims to verify the performance of the multiobjective collaborative optimization algorithm for image matting based on fuzzy multicriteria evaluation and decomposition. The experiment was conducted not only on 27 images with ground-truth alpha mattes in the benchmark dataset but also on the remaining eight images without publicly available ground-truth alpha mattes through online testing. The experiment on the 27 images used two types of trimaps, one with a large unknown region and the other with a small unknown region. The experiment on the eight images used three different types of trimaps: a large unknown region, a small unknown region, and a manually calibrated one. This experiment used three widely used matting performance indicators, including the absolute error sum, mean square error, and gradient error, to quantitatively evaluate the matting performance of existing algorithms. The experiment used three advanced sampling-based image matting algorithms as performance benchmarks. The algorithms used were the KL-divergence sampling-based image matting algorithm [7], the comprehensive sampling-based image matting algorithm [6], and the weighted color and texture image matting algorithm [5]. Considering the impact of preprocessing and postprocessing on segmentation performance, all methods in the experiment did not use preprocessing or postprocessing to ensure fairness. It is worth noting that the KL-divergence sampling-based image matting algorithm failed to collect background pixels during the segmentation process of the GT02 and GT25 images because all collected pixels fell within the known foreground area. Therefore, the alpha matte for these two images could not be obtained for this algorithm.

Tables 1.8 and 1.9 show the performance of the multiobjective collaborative optimization image matting algorithm based on fuzzy multicriteria evaluation and decomposition, and three advanced sampling-based image matting algorithms on the mean squared error of alpha mattes for 27 images and two types of corresponding trimaps. In the experiment with larger unknown regions of the trimap, the average mean squared errors of the multiobjective collaborative optimization image matting algorithm, KL-divergence-based sampling image matting algorithm, comprehensive sampling image matting algorithm, and weighted color and texture image matting algorithm are respectively 538.5, 559.3, 755.4, and 666.8 for 27 images. In the experiment with smaller unknown regions of the trimap, the average mean squared errors of the introduced algorithm, KL-divergence-based sampling image matting algorithm, comprehensive sampling image matting algorithm, and weighted color and texture image matting algorithm are respectively 292.0, 410.1, 547.1, and 460.4 for 27 images. The mean squared error values of the multiobjective collaborative optimization image matting algorithm are larger than those of the other three advanced image matting algorithms on both types of trimaps with smaller and larger unknown regions. The multiobjective collaborative optimization image matting algorithm not only performs better than existing image matting algorithms in terms of mean squared error, but also achieves lower mean squared error in most images. In the comparison of results with larger unknown regions of the trimap, the multiobjective collaborative optimization image matting algorithm obtained the lowest mean squared error in 21 out of 27 images. In the comparison of results with smaller unknown regions of the trimap, the multiobjective collaborative optimization image matting algorithm obtained the lowest mean squared error in 22 images. Table 1.10 presents the results of online image matting benchmark tests, detailing the absolute error sum, mean squared error (MSE), and gradient error for a multi-objective collaborative optimization image matting algorithm based on fuzzy multi-criteria evaluation and decomposition, alongside three other advanced sampling-based image matting algorithms. Multiobjective collaborative optimization image matting algorithm outperforms existing image matting algorithms in all three alpha matte performance indicators on different types of trimaps. These experimental results demonstrate that multi-objective collaborative optimization image matting algorithm has better image matting performance than existing algorithms.

Fig. 1.12 provides a comparison of alpha mattes obtained by the multiobjective collaborative optimization algorithm based on fuzzy multicriteria

Table 1.8 The mean square error (MSE) of the alpha matte produced by a multi-objective collaborative optimization matting algorithm based on fuzzy multi-criteria evaluation and decomposition is compared across 27 images with that of existing image matting algorithms. The trimaps with small unknown regions were employed.

Image name	GT01	GT02	GT03	GT04	GT05	GT06	GT07	GT08	GT09
Multiobjective collaborative optimization image matting algorithm	**29.4**	**67.3**	**125.1**	390.6	**36.9**	**71.8**	**37.1**	**613.8**	**99.3**
KL-divergence matting [7]	48.7	N/A	157.6	**282.7**	133.8	144.2	61.2	715.5	141.4
Comprehensive sampling matting [6]	56.9	119.8	170.1	499.8	137.3	173.6	60.9	782.9	114.3
Weighted color and texture matting [5]	39.4	169.5	126.9	340.6	100.8	138.5	58.3	711.3	228.8

Image name	GT10	GT11	GT12	GT13	GT14	GT15	GT16	GT17	GT18
Multiobjective collaborative optimization image matting algorithm	**149.7**	**213.3**	**59.3**	**228.3**	**63.1**	**181.2**	**927.0**	**58.6**	**44**
KL-divergence matting [7]	244.9	264.6	101.8	334.1	126.0	292.8	3084.8	109.1	86.0
Comprehensive sampling matting [6]	287.5	236.9	106.4	414.9	120.5	249.7	4553.1	194.9	114.8
Weighted color and texture matting [5]	183.7	319.7	79.8	363.1	153.7	279.2	4407.8	157.6	82.4

Image name	GT19	GT20	GT21	GT22	GT23	GT24	GT25	GT26	GT27
Multiobjective collaborative optimization image matting algorithm	**76.3**	**50.4**	349.6	**40.8**	**50.7**	**359.7**	1245.8	1003.0	1311.0
KL-divergence matting [7]	91.5	113.7	**311.2**	86.7	67.2	607.3	N/A	1415.0	1231.4
Comprehensive sampling matting [6]	160.8	114.9	587.0	85.7	117.1	432.6	1721.4	1013.4	2085.3
Weighted color and texture matting [5]	93.2	129.2	589	71.8	78.4	492.2	**997.5**	**962.5**	**1075.8**

KL, Kullback–Leibler.

Note: The best results are highlighted in bold.

Table 1.9 The mean square error (MSE) of the alpha matte produced by a multi-objective collaborative optimization matting algorithm based on fuzzy multi-criteria evaluation and decomposition is compared across 27 images with that of existing image matting algorithms. The trimaps with large unknown regions were employed.

Image name	GT01	GT02	GT03	GT04	GT05	GT06	GT07	GT08	GT09
Multiobjective collaborative optimization image matting algorithm	**58.8**	**143.1**	206.5	673.5	**89.2**	141.1	54.7	774.2	**116.9**
KL–divergence matting [7]	73.7	143.4	188.2	**378.9**	203.7	295	73.8	925.8	189.4
Comprehensive sampling matting [6]	100.5	197.3	233.1	691.4	213.2	354.6	100.0	948.7	174.0
Weighted color and texture matting [5]	64.9	244.2	**135.9**	531.6	155.3	203.1	87.8	883.6	246.6

Image name	GT10	GT11	GT12	GT13	GT14	GT15	GT16	GT17	GT18
Multiobjective collaborative optimization image matting algorithm	240.0	381.3	99.9	463.1	114.1	299.3	2282.3	90.2	89.9
KL–divergence matting	349.2	531.3	157.9	481.3	192.6	514.4	4423.0	329	195.7
Comprehensive sampling matting	541.9	471.9	135.5	673.1	256.4	420.8	4583.9	211.5	213.3
Weighted color and texture matting	279.8	401.3	124.8	754.0	239.8	548.8	4578.3	209.0	187.3

Image name	GT19	GT20	GT21	GT22	GT23	GT24	GT25	GT26	GT27
Multiobjective collaborative optimization image matting algorithm	**199.6**	110.8	826.4	74.9	87.7	648.6	1786.2	1689.5	2797.3
KL–divergence matting [7]	229.4	214.0	**604.3**	217.8	128.1	854.4	N/A	**1415.0**	**1231.4**
Comprehensive sampling matting [6]	355.5	199.4	1152.4	213.0	174.3	777.8	2164.0	1619.2	3218.3
Weighted color and texture matting [5]	271.2	281.5	1053.2	101.5	133.3	741.2	**1724.5**	1757.7	2063.2

KL, Kullback–Leibler.
Note: The best results are highlighted in bold.

Table 1.10 Multiobjective collaborative optimization image matting algorithm based on fuzzy multicriteria evaluation, in comparison to decomposition and online image matting benchmarking results of three advanced sampling-based image matting algorithms.

Sum of absolute error	Overall rank	Avg. small rank	Avg. large rank	Avg. user rank
Multiobjective collaborative optimization image matting algorithm	**18.3**	**15.8**	**21.9**	**17.4**
KL–divergence matting [7]	18.9	16.6	22.1	18.1
Comprehensive sampling matting [6]	20.1	17.7	23.5	19.0
Weighted color and texture matting [5]	21.2	19.3	24.5	19.7
Mean square error	Overall rank	Avg. small rank	Avg. large rank	Avg. user rank
Multiobjective collaborative optimization image matting algorithm	**1.4**	**1.1**	**1.7**	**1.3**
KL–divergence matting [7]	1.5	1.3	1.8	1.4
Comprehensive sampling matting [6]	1.6	1.4	2.0	1.5
Weighted color and texture matting [5]	1.8	1.7	2.2	1.7
Gradient error	Overall rank	Avg. small rank	Avg. large rank	Avg. user rank
Multiobjective collaborative optimization image matting algorithm	**1.2**	**1.0**	**1.3**	**1.1**
KL–divergence matting [7]	1.4	1.3	1.5	1.4
Comprehensive sampling matting [6]	1.5	1.3	1.6	1.4
Weighted color and texture matting [5]	2.0	1.9	2.2	2.0

KL, Kullback–Leibler.

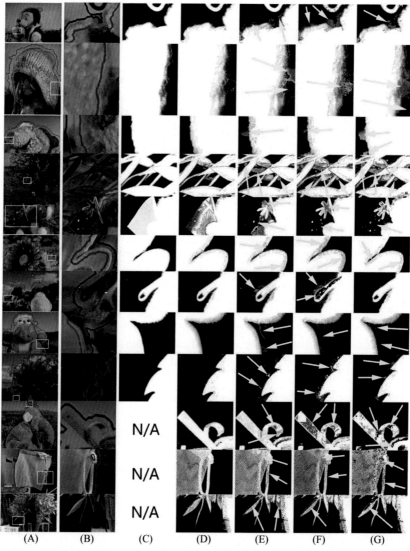

N/A

N/A

N/A

(A)   (B)   (C)   (D)   (E)   (F)   (G)

Figure 1.12 Comparison of alpha mattes obtained by multiobjective collaborative optimization image matting algorithm based on fuzzy multicriteria evaluation and decomposition with existing image matting algorithms. (A) Images with boundaries of known foreground regions marked in red and boundaries of known background regions marked in blue and zoomed windows marked in yellow. (B) Zoomed-in region. (C) Ground-truth alpha matte. (D) the introduced algorithm. (E) KL-Divergence matting [7]. (F) Comprehensive sampling matting [6]. (G) Weighted color and texture matting [5]. Arrows indicate regions of low-quality mattes. *KL*, Kullback-Leibler.

evaluation and decomposition and existing algorithms involved in the experiment. Fig. 1.12A is the input image; Fig. 1.12B is a zoomed image of the yellow region in (A); Fig. 1.12C is the ground-truth alpha matte; Fig. 1.12D is the alpha matte obtained by the multiobjective collaborative optimization algorithm based on fuzzy multicriteria evaluation and decomposition; Fig. 1.12E is the alpha matte obtained by the KL-divergence sampling-based image matting algorithm; Fig. 1.12F is the alpha matte obtained by the comprehensive sampling-based image matting algorithm; and Fig. 1.12G is the alpha matte obtained by the color and texture feature weighted image matting algorithm. It can be observed from the figure that the multiobjective collaborative optimization algorithm based on fuzzy multicriteria evaluation and decomposition can achieve better segmentation results. The algorithm-generated alpha matte for image matting has lower noise and clearer edges, visually resembling the ground-truth alpha matte for image matting. The visual contrast results of the alpha matte are consistent with the quantitative comparison results, which further indicates that the multiobjective collaborative optimization image matting algorithm can provide high-quality alpha matte.

The improvement in image matting performance brought about by the introduced multiobjective collaborative optimization image matting algorithm based on fuzzy multicriteria evaluation and decomposition can be attributed to three factors:

1) The fuzzy multicriteria pixel pair evaluation method effectively deals with the uncertainty of multiple evaluation criteria and provides an accurate objective function for pixel pair optimization.

2) The multicriteria evaluation function decomposition optimization strategy effectively utilizes the additional heuristic information provided by each criterion by transforming the multicriteria single-objective optimization problem into multiple single-objective optimization problems, thereby improving the global search ability of heuristic optimization algorithms.

3) Based on neighborhood grouping collaborative multiobjective optimization strategy and taking advantage of local image smoothness assumptions, the large-scale pixel pairing optimization problem is quickly and accurately decomposed into smaller subproblems, thus improving the search efficiency of the pixel pairing optimization problem.

However, the multiobjective collaborative optimization algorithm based on fuzzy multicriterion evaluation and decomposition still has some shortcomings. This method uses low color distortion and spatial distance similarity evaluation criteria. In texture-rich images, this evaluation method may have evaluation errors, leading to a decrease in the quality of the obtained alpha matte. Therefore, the multiobjective collaborative optimization algorithm based on fuzzy multicriterion evaluation and decomposition exhibits a large mean square error in texture-rich images such as GT25, GT26, and GT27, indicating that its performance on these images is not as good as algorithms that consider texture features (as shown in Tables 1.8 and 1.9). This experimental result shows that texture similarity evaluation criteria can effectively distinguish high-quality pixel pairs in these examples, whereas color distortion and spatial criteria cannot effectively distinguish high-quality pixel pairs. Although the performance of the multiobjective collaborative optimization image matting algorithm is not the best in these examples, thanks to the effective handling of the uncertainty of the fuzzy multicriterion pixel pair evaluation criteria, its matting performance is still competitive compared to other algorithms that do not consider texture similarity evaluation criteria.

1.2.3 Summary

This section focuses on the problem of low optimization accuracy faced by nonsampling image matting based on heuristic optimization and introduces how to use the characteristics of pixel pair optimization problems to further improve the solution accuracy of heuristic optimization-based image matting algorithms. To address the issue of uncertain satisfaction levels of multiple evaluation criteria in pixel pair evaluation, a fuzzy multicriteria pixel pair evaluation method is introduced based on fuzzy logic. To fully utilize the heuristic information provided by multiple evaluation criteria and use the prior knowledge of evaluation criteria, a decomposition-based multiobjective collaborative optimization algorithm is introduced in this section. In this algorithm, the multicriteria evaluation function decomposition optimization strategy is introduced to decompose the multicriteria single-objective optimization problem into multiple single-objective optimization problems, and a multiobjective optimization algorithm is used to optimize all optimization problems simultaneously. In addition, the multiobjective collaborative optimization image matting algorithm uses the characteristic of local image smoothing and employs a neighborhood grouping collaborative multiobjective optimization strategy. This strategy decomposes the large-scale pixel pair optimization problem into multiple small-scale optimization problems and optimizes them independently.

Experimental results show that the fuzzy multicriteria pixel pair evaluation method not only provides accurate evaluations in the average sense but also accurately evaluates pixel pairs even when some evaluation criteria are not satisfied to a high degree. In the experiment, we found that using the multicriteria evaluation function decomposition optimization strategy can transform the complex multicriteria single-objective pixel pair optimization problem into a multiobjective optimization problem, which can help us obtain higher solution accuracy than directly performing single-objective optimization. In addition, a fast and effective problem decomposition method for large-scale pixel pair optimization is provided, based on neighborhood grouping collaborative multiobjective optimization strategy, which effectively improves the optimization accuracy of heuristic optimization algorithms. The experiment also shows that the multiobjective collaborative optimization algorithm based on fuzzy multicriteria evaluation and decomposition can provide high-quality alpha mattes for different types of images and different trimap situations. Its performance in terms of absolute error, mean square error, and gradient error is better than existing advanced image matting algorithms. However, the limitation of this algorithm is that it does not consider texture features. Therefore, there is a performance decrease in image matting for texture-rich images. Also, the high computational complexity due to its use of the heuristic optimization algorithm in this algorithm limits its application in tasks with high time constraints. Future research should focus on improving the speed of image matting while ensuring accuracy.

1.3 Another masterpiece when medicine meets artificial intelligence

1.3.1 Overview of research progress

In current medical practice, vessel segmentation technology plays an important role in retinal image analysis and computer-aided diagnosis of eye diseases. It is the basis for medical diagnosis and surgical assistance design. It is also of great significance for the early detection and treatment of different cardiovascular and eye diseases such as stroke, venous obstruction, diabetic retinopathy, and arteriosclerosis. In recent years, vessel segmentation has become one of the hot issues in the field of medical image processing, and many automatic vessel segmentation techniques have been introduced with good segmentation results. However, the use of matting as an auxiliary technique in vessel segmentation is rare. So far, only one patent has been found that uses invariant moment features and KNN matting algorithm [13] for vessel segmentation. However, since generating a trimap is a tedious and time-consuming task in vessel segmentation, it is currently necessary to design a matting algorithm to segment vessels as efficiently as possible.

1.3.2 Scientific principles

1.3.2.1 Blood vessel segmentation algorithm based on hierarchical matting model

As shown in Fig. 1.13, the blood vessel segmentation algorithm based on the hierarchical matting model consists of two steps: trimap generation and matting.

The automatic process of generating a vascular trifurcation diagram is shown in Fig. 1.14, which mainly includes two steps: image matting and extraction of vascular skeletons.

1.3.2.1.1 Image segmentation

Fig. 1.15 shows a schematic diagram of the process effect of image segmentation. Firstly, the image is segmented into three regions: background region B, unknown region U, and initial vessel region V_1. The segmentation method is shown in Eq. (1.23).

$$I_{mr} = \begin{cases} B & \text{if } 0 < I_{mr} < p_1 \\ U & \text{if } p_1 \leq I_{mr} < p_2 \\ V_1 & \text{if } p_2 \leq I_{mr} \end{cases} \tag{1.23}$$

In order to minimize the unknown region, the value of p_1 is set to 0.2, and the value of p_2 is set to 0.35 [6,30]. Then the characteristics of the vascular area are used to remove the noise area in V_1. Vascular region characteristics have been widely used in vascular segmentation, achieving good

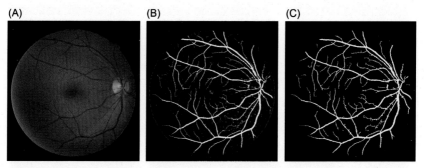

Figure 1.13 Flowchart of blood vessel segmentation algorithm based on hierarchical image matting model. Automatic generation of vascular trifurcation map.

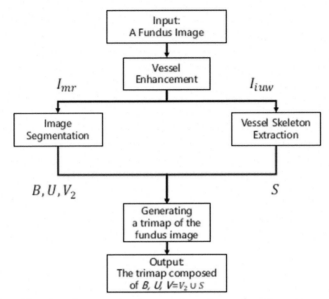

Figure 1.14 Flowchart for automated generation of vascular trifurcation map.

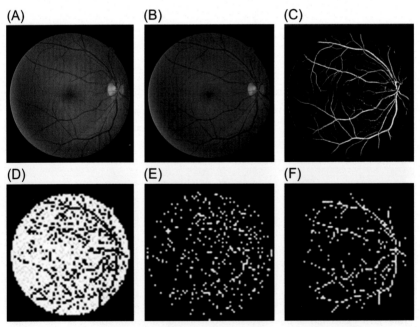

Figure 1.15 Schematic diagram of blood vessel image segmentation effect.

segmentation accuracy and computational efficiency [31]. Region features describe the vascular area through area, external rectangle, and convex hull information. The denoised image is denoted as V_2.

1.3.2.1.2 Extraction of vascular skeleton

The goal of vessel extraction is to further distinguish unknown regions and provide more vascular information. The first step is to obtain a binary image T through binarization of the input image by Eq. (1.24).

$$T = \begin{cases} 1 & I_{iuw} > t \\ 0 & I_{iuw} \leq t \end{cases} \tag{1.24}$$

where $t = Otsu(I_{iuw}) - \varepsilon$, $Otsu$ denotes the Otsu's thresholding algorithm and ε is set to 0.03. In the second step, T is divided into three parts (T_1, T_2, and T_3) according to the number of pixels in the region (Area).

$$T = \begin{cases} T_1 & if \ 0 < \ Area \ < a_1 \\ T_2 & if \ a_1 \leq Area \leq a_2 \\ T_3 & if \ a_2 < \ Area \end{cases} \tag{1.25}$$

where $a_1 = 42 \times \frac{max(h,w)}{min(h,w)}$, $a_2 = 735 \times \frac{max(h,w)}{min(h,w)} \cdot h$ and w denote the height and width of the image.

During the process of extracting the vascular skeleton, the T_1 area is removed and T_3 is retained. Then T_2 is further subdivided into T_4 according to the characteristics of the vascular region. Finally, T_3 and T_4 regions are combined, and the vascular skeleton is obtained using a skeleton extraction algorithm [32]. Fig. 1.16 shows an example of vascular skeleton extraction.

After completing the steps of image segmentation and blood vessel skeleton extraction, the blood vessel area $V = V_2 \cup S$, the background area B, and the unknown area U can all be obtained in the trimap.

1.3.2.1.3 Hierarchical matting model

The hierarchical matting model incrementally labels the unknown regions as blood vessels or background regions. If the pixels in the unknown region are divided into m layers and the ith pixel in the unknown region belongs to the j-th layer, the blood vessel image can be represented as follows:

$$I_v(z) = \begin{cases} 1 & if \ corre \ (u_i^j, V) > corre \ (u_i^j, B) \\ 0 & else \end{cases} \tag{1.26}$$

corre represents the similarity function [see Eq. (1.29)].

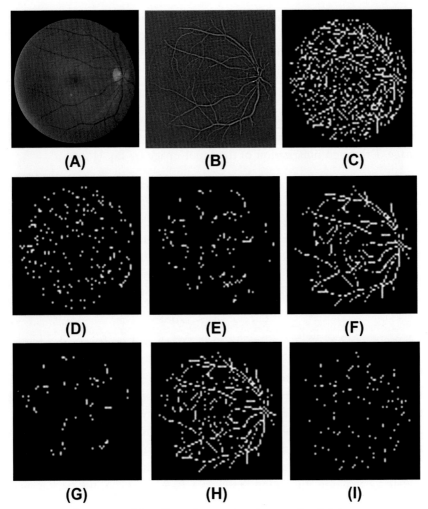

Figure 1.16 Illustration of the effect of extracting the vascular skeleton.

The implementation of the hierarchical segmentation model consists of the following two steps.

1) Dividing unknown pixels into layers: Assign unknown area pixels to different layers.
2) Layer-wise updating: Allocate new labels (vessel or background) for each layer of pixels. The pseudocode for the layered matting model is shown in Algorithm 1.4.

ALGORITHM 1.4 Hierarchical matting algorithm
Input: Trimap composed of B, B,U,V
Output: The segmented vessel image I_u
  Step 1: Stratifying the unknown
  pixels:
1: For $i = \{1,2,\ldots,n_U\}$, set $D(i) = d_i$, where n_U is the number of
  unknown pixels in U, d_i is the distance between the ith unknown
  pixel and the closest vessel pixel in V, D is the set of d_i
2: Sort the unknown pixels in U in ascending order according to the
  distances D, cluster the pixels with the same distance into one hierarchy,
  stratify the pixels into m hierarchies
  and denote them as an hierarchical order set: $H = \{H_1,H_2,\ldots,H_m\}$,
  $H_j = u^j_i | i \in 1,2,\ldots,n_i$, where n_i is the number of unknown pixels in
  the j-th hierarchy.
  Step 2: Hierarchical update
3: **for** $j = 1,\ldots,m$ **do**
4:     **for** $i = 1,\ldots,n_i$ **do**
5: Compute the correlations [defined in Eq. (1.29)] between uj i and its
  neighboring labeled pixels(vessel pixels and background pixels)
  included in a 9×9 grid.
6:     Choose the labeled pixel with the closest correlation, and assign
      its label to u^j_i
7:   **end**
for
8: **end for**

1.3.2.1.4 Stratifying the unknown pixels

For the ith pixel in an unknown area, the distance is calculated between it and all pixels in the vascular area V, and the shortest distance is assigned to the ith pixel. Classify the unknown pixels based on their shortest distance, with the pixel with the shortest distance assigned to the first layer and the pixel with the longest distance assigned to the last layer. An unknown pixel located in the first layer indicates its proximity to the blood vessels, while an unknown pixel located in the last layer indicates its distance from the blood vessels. Fig. 1.17 shows an example of unknown pixel classification.

1.3.2.1.5 Correlation function

In step 2 of Algorithm 1.4, given an unknown pixel u^j_i and its neighboring labeled pixel k^j_i, the color cost function used to describe the fitness between u^j_i and k^j_i can be defined as follows:

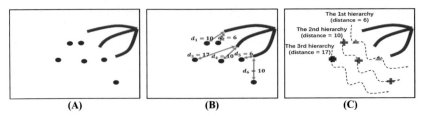

Figure 1.17 Pseudocode for hierarchical image matting model.

$$\beta_c\left(u_i^j, k_j^i\right) = \left\| c_{u_i^j} - c_{k_i^j} \right\| \tag{1.27}$$

$c_{u_j}i$ and $c_{k}lj$ refer to the intensity of u_i^j and k_j^i in $I_m r$, respectively. The spatial cost function is defined as follows:

$$\beta_s\left(u_i^j, k_l^j\right) = \frac{\left\| x_{u_i^j} - x_{k_l^j} \right\| - x_{\min}}{x_{\max} - x_{\min}} \tag{1.28}$$

where $x_{u_i^j}$ and $x_{k_l^j}$ are the spatial coordinates of u_i^j and k_l^j, $x_{\max}$ and $x_{\min}$ represent the longest and shortest distance from the unknown pixel u_i^j to the labeled pixel k_l^j, respectively. Normalization factors $x_{\max}$ and $x_{\min}$ ensure that β_s is independent of absolute distance.

The final similarity function β is a weighted combination of color fitness and spatial distance:

$$\beta\left(u_i^j, k_l^j\right) = \beta_c\left(u_i^j, k_l^j\right) + \omega\beta_s\left(u_i^j, k_l^j\right) \tag{1.29}$$

where ω is the weighting factor for balancing color adaptation and spatial distance, set to 0.5. A smaller ω layer by layer update.

1.3.2.1.6 Hierarchical update
After the initialization of the layering strategy is completed, the similarity between each unknown pixel and the labeled pixels within its 9×9 neighborhood is calculated, and the label of the labeled pixel with the strongest similarity is assigned to the unknown pixel. Once all unknown pixels in a layer have been updated, the pixel information from that layer is used to update the next layer of pixels. This process is repeated from the first layer to the last, as shown in Fig. 1.18.

1.3.2.2 Experimental analysis
The effectiveness of the hierarchical matting algorithm has been verified through comparison with multiple blood vessel segmentation algorithms. The comparison results are shown in Table 1.11. The experimental results

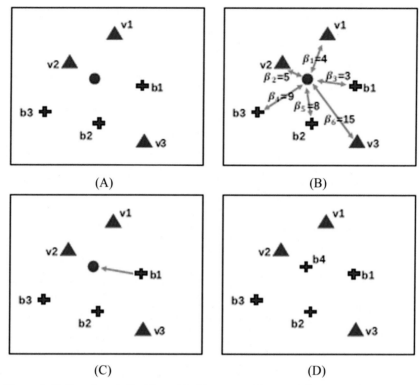

Figure 1.18 Pseudocode for hierarchical image matting model.

show that the blood vessel segmentation algorithm based on the hierarchical masking model achieves good segmentation results on both the publicly available databases DRIVE [33] and STARE. Although the supervised method proposed by Hoover et al. [34] performs best on the STARE database, it results in high computational cost due to its use of deep learning techniques. Compared with other algorithms, the introduced method has the advantage of lower computational cost.

Fig. 1.19 shows some examples of blood vessel segmentation comparison results. The left image in Fig. 1.19 is the optimal segmentation result, and the right image is the result obtained based on the algorithm. It can be seen that the segmentation result obtained by the algorithm extracted more complex feature information compared to that of the doctor's segmentation result.

1.3.3 Summary

In this section, the problem of automatic generation of trimap is explored using blood vessel image matting as an example. Image enhancement and

Table 1.11 Comparison of the segmentation performance of the introduced algorithm with the world's leading algorithms.

Test datasets	DRIVE					STARE					
Methods	Ace	AUC	Se	Sp	Time	Acc	AUC	Se	Sp	Time	System
Supervised methods											
Staal et al.	0.944	—	—	—	15 min	0.952	—	—	—	15 min	1.0 GHz, 1-GB RAM
Soares et al.	0.946	—	—	—	~3 min	0.948	—	—	—	~3 min	2.17 GHz, 1-GB RAM
Lupascu et al.	0.959	—	0.720	—	—	—	—	—	—	—	—
Marin et al.	0.945	0.843	0.706	0.980	~90 s	0.952	0.838	0.694	0.982	~90 s	2.13 GHz, 2-GB RAM
Roychowdhury et al.	0.952	0.844	0.725	0.962	3.11 s	0.951	0.873	0.772	0.973	6.7 s	2.6 GHz, 2-GB RAM
Liskowski et al.	0.954	0.881	0.781	0.981	—	0.973	0.921	0.855	0.986	—	NVIDIA GTX Tian GPU
Unsupervised methods											
Hoover et al.	—	—	—	—	—	0.928	0.730	0.650	0.810	5 min	Sun SPARC station 20
Mendonca et al.	0.945	0.855	0.734	0.976	2.5 min	0.944	0.836	0.699	0.973	3 min	3.2 GHz, 980-MB RAM
Lam et al.	—	—	—	—	—	0.947	—	—	—	8 min	1.83 GHz, 2-GB RAM
Al-Diri et al.	—	—	0.728	0.955	11 min	—	—	0.752	0.968	—	1.2 GHz
Lam and Yan	0.947	—	—	—	13 min	0.957	—	—	—	13 min	1.83 GHz, 2-GB RAM
Perez et al.	0.925	0.806	0.644	0.967	~2 min	0.926	0.857	0.769	0.944	~2 min	Parallel Cluster
Miri et al.	0.943	0.846	0.715	0.976	~50 s	—	—	—	—	—	3 GHz, 1-GB RAM
Budai et al.	0.957	0.816	0.644	0.987	—	0.938	0.781	0.580	0.982	—	2.3 GHz, 4-GB RAM
Nguyen et al.	0.941	—	—	—	2.5 s	0.932	—	—	—	2.5 s	2.4 GHz, 2-GB RAM
Yitian et al.	0.954	0.862	0.742	0.982	—	0.956	0.874	0.780	0.978	—	3.1 GHz, 8-GB RAM
Annunziata et al.	—	—	—	—	—	0.956	0.849	0.713	0.984	<25 s	1.9 GHz, 6-GB RAM
Orlando et al.	—	0.879	0.790	0.968	—	—	0.871	0.768	0.974	—	2.9 GHz, 64-GB RAM
Proposed	0.960	0.858	0.736	0.981	10.72 s	0.957	0.880	0.791	0.970	15.74 s	2.5 GHz, 4-GB RAM

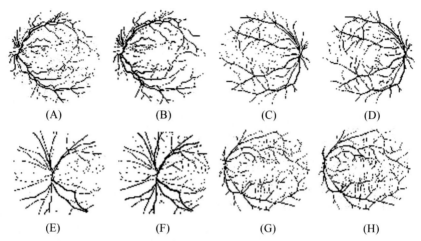

Figure 1.19 Visualization comparison of vascular segmentation effects.

the shape features of blood vessels are used to achieve the automatic generation of trimap, and the algorithm based on a hierarchical matting model for blood vessel image matting extends the application of natural image matting technology in the field of medical imaging. This technology is expected to be applied in scenarios such as the segmentation of complex objects and processing of small-sample data. In future work, we will continue to optimize this algorithm, improve its performance, and collaborate with experts in the medical field to apply the algorithm to practical assisted diagnosis as much as possible.

1.4 "Deep learning + image matting enhancement" trial with great effectiveness: pedestrian classification in infrared images

1.4.1 Overview of research progress

Pedestrian classification is an important research topic in computer vision with significant theoretical and practical value. Infrared images, due to their unique advantages over visible light images—being unaffected by weather and lighting conditions, have received extensive attention from researchers in pedestrian classification. Infrared image pedestrian classification can provide key technical support for advanced driver assistance systems. Fig. 1.20 shows the structure diagram of an advanced driver assistance system. It can be seen from the figure that robust pedestrian classification is an important part of the advanced driver assistance system,

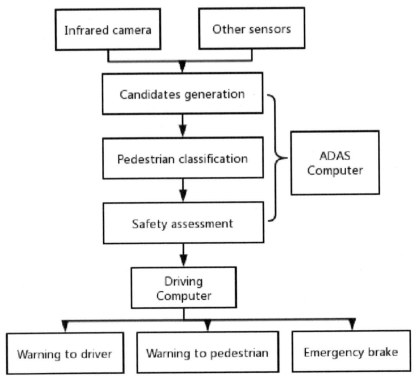

Figure 1.20 Block diagram of the architecture of the advanced driver assistance system.

and the pedestrian classification result is an important basis for driving safety assessment. In practical applications, pedestrian classification errors may lead to serious traffic accidents.

Although far-infrared images have unique advantages in pedestrian classification, they have the problems of low resolution and lack of color information, resulting in extremely limited information contained in far-infrared images. Therefore, high-resolution features such as texture commonly used in visible light images cannot be applied to far-infrared images. As far-infrared images are based on temperature differences, the pixel intensity values of pedestrian areas in far-infrared images may change with the seasons and temperature. For example, in far-infrared images taken in winter, the pedestrian areas are relatively brighter than the background; while in summer, the pedestrian areas may be darker than the background [35]. The pixel intensity values of the background in far-infrared images may be very close to those of the pedestrian areas, which poses great challenges for infrared image pedestrian classification.

The core of infrared image pedestrian classification is to extract the representation form of pedestrians. Its research process can be divided into two stages: expert-driven algorithms and data-driven algorithms.

Expert-driven far-infrared image pedestrian classification algorithms are designed based on the experiential knowledge of experts in the pedestrian classification field. These algorithms typically involve two steps: feature extraction and pedestrian classification. Feature extraction is used to transform the original image into a representation that has been carefully designed by domain experts (such as a feature vector), which allows the classifier to effectively distinguish between pedestrian and nonpedestrian targets in this representation. The design of the representation relies on expert knowledge in the far-infrared image pedestrian classification field. As color information is not available in far-infrared images, the contour features of pedestrians become an important feature for pedestrian classification. To design high-quality features that not only have separability for representing pedestrian information but also remain invariant to irrelevant cluttered backgrounds, researchers have introduced different contour features [35−40].

Expert-driven far-infrared image pedestrian classification algorithms can be further divided into two categories: histogram-based algorithms and nonhistogram-based algorithms. Dalal et al. [41] found that the magnitude and direction of gradients are robust image features and designed oriented gradient histogram features for pedestrian detection. Suard et al. [38] extended the gradient direction histogram to pedestrian detection in infrared images. Zhao et al. [35] introduced a shape distribution histogram based on contour maps. The histogram measures the distance between any two points selected arbitrarily on the contour line in the candidate area to describe the shape of the object. Histogram-based algorithms can tolerate small changes in pedestrian shape. However, when the brightness of the pedestrian area is similar to the cluttered background, these algorithms cannot effectively represent indistinct contours, resulting in incorrect pedestrian discrimination. Therefore, recent researchers have introduced nonhistogram-based algorithms. Bassem et al. [37] introduced a far-infrared pedestrian classification algorithm based on SURF features [42]. This algorithm assumes that the head of the pedestrian in the far-infrared image is brighter than the background and describes the key features of the pedestrian head by learning hierarchical codebooks on the SURF features. Although this algorithm is not affected by the motion posture of the pedestrian, it ignores the information of the upper body of pedestrians, so its robustness cannot be guaranteed under complex interference. Kwak et al. [43] introduced a far-infrared image pedestrian classification algorithm based on oriented center symmetric local binary patterns, assuming that the pedestrian area in the far-infrared image is brighter than the background. The algorithm divides the region of interest (ROI) into nonoverlapping 4×4 image blocks and uses rotation-invariant oriented center symmetric local binary pattern features to describe the brightness information of each image block. By concatenating the feature vectors of each pixel block, a 128-dimensional feature vector is formed, and the representation of the pedestrian is learned using a cascaded random forest classifier. The oriented center symmetric local binary pattern features used by this algorithm solve the problem of poor robustness of histogram-based algorithms under interference. However, this algorithm cannot provide accurate classification results when the motion amplitude of the pedestrian is large.

The expert-driven algorithms are based on empirical models of local luminance designs for infrared pedestrians. With the season, temperature, and even background changes of pedestrians, the brightness characteristics of pedestrians in far-infrared images will change greatly. Therefore, expert-driven algorithms have poor classification performance in complex situations such as similar pedestrians and backgrounds.

With the development of deep learning techniques, data-driven pedestrian classification algorithms have received increasing attention. Data-driven pedestrian classification algorithms, represented by deep learning techniques, can automatically learn representational information from the data without manual intervention [44]. By simulating the information processed by biological neural systems, such algorithms use multilayer convolutional neural networks to find data structure features that are difficult to find from high-dimensional data. Data-driven algorithms can solve different types of problems through universal learning programs, and have made major breakthroughs in [45–47] numerous computer vision tasks, including image recognition.

Increasing the number of layers of convolutional neural networks is considered an effective way to improve the performance of data-driven classification algorithms. To deal with complex examples (such as pedestrian background interference in far-infrared images), the researchers designed deep neural networks with deeper layers.

Krizhevsky et al. [45] proposed a neural network called AlexNet, where the first five layers are convolutional layers and the remaining three layers are fully connected layers. AlexNet uses the SigMoid function to replace the linear rectification function (rectified linear unit) as the activation function, giving it a higher convergence rate in the stochastic gradient descent optimization. Simonyan et al. [46] conducted a rigorous evaluation of the neural networks with different layers, and experimentally verified that the deeper the number of the neural network layers, the better the performance. Therefore, the research team raised the number of layers of the neural network to 16−19 layers and proposed a neural network called VGG. By stacking small 3×3 convolution filters, the VGG network managed to reduce the number of parameters and maintain the visual receptive field. However, VGG networks still consume a lot of time and memory in the prediction. The ResNet proposed by He et al. [47] increases the number of neural network layers to 152 layers. By skipping one or more layers and introducing stimuli from the front layer through short connections, the network alleviates the training gradient disappearance problem that occurs when the neural network layers are too deep providing a feasible solution to the problem of high training error of neural networks with very deep layers. ResNet does not solve the gradient vanishing problem but avoids the problem through the combination of very shallow networks. These increasingly deep neural networks set broken records on benchmarks for multiple computer vision tasks, including pedestrian classification. The computational time and spatial complexity of the deep neural network also increase with the number of network layers, and the deep number of layers of the neural network limits its practical application. The solution based on GPU acceleration can reduce the computing time of deep neural networks. However, considering that the device in practical applications often cannot be equipped with GPUs, the computational cost of neural networks with deep network layers is too high for industrial applications.

In summary, it is necessary to study how to improve the classification performance of deep neural networks at a lower cost. Considering the significant impact of cluttered backgrounds on the effective extraction of pedestrian contour features in infrared image pedestrian classification, which leads to a decrease in classification accuracy, it is necessary to accurately separate infrared targets from complex backgrounds to effectively extract contour features. Natural image matting technology provides a tool for accurately separating foregrounds from cluttered backgrounds; however, the time-consuming nature of manually labeled trimaps that this technology relies on makes it not viable for practical application. Therefore, it is necessary to study an automatic infrared image pedestrian image matting algorithm that does not rely on manually labeled trimaps.

1.4.2 Scientific principles

1.4.2.1 Pedestrian classification algorithm based on automatic image matting and enhancement for infrared images

This section introduces a pedestrian classification algorithm for infrared images based on automatic image matting enhancement to address the problem of cluttered backgrounds [48]. Fig. 1.21 shows this algorithm structure

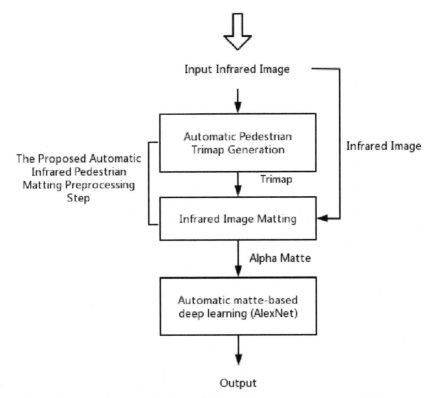

Figure 1.21 Structural block diagram of infrared image pedestrian classification algorithm based on automatic image matting enhancement.

diagram. The algorithm organically combines expert-driven algorithms with data-driven algorithms, mainly including infrared pedestrian preprocessing based on automatic image matting and deep pedestrian classification based on alpha mattes.

1) By using the automatic infrared image pedestrian segmentation algorithm, image enhancement preprocessing of far-infrared images has been achieved. The messy background in the alpha matte of the image after preprocessing has been effectively suppressed and the pedestrian area has been enhanced, allowing deep neural networks to better learn pedestrian contour features.

2) The preprocessed alpha matte from the clipped image is utilized as input for the AlexNet classifier. Algorithm 1.5 provides an implementation process of the infrared pedestrian classification algorithm (Algorithm 1.5).

ALGORITHM 1.5 Infrared pedestrian classification algorithm based on automatic image matting enhancement

Input: the region of interest in the far-infrared image I

output: Whether pedestrians are included in I

1: $\tilde{I} \leftarrow standardizeImage(I)$

    //1. Infrared pedestrian preprocessing based on automatic image matting

    // 1.1 Generate infrared pedestrian trimap automatically

2: the head position $\leftarrow locateHead(\tilde{I})$

3: the body position $\leftarrow locateBody(\tilde{I}$, the head position)

4: trimap $\leftarrow generateTrimap($the head position, the body position)

    //1.2 Far-infrared image matting

5: Generate candidates by global sampling

6: **for each** unknown pixel **do**

7:     **for each** *candidate foreground/background pixel pairs* **do**

8:       Evaluation of the foreground/background pixel pairs

9:     **end for**

10:    Foreground/background pixel pair selection

11:    Estimation of alpha values

12: **end for**

    // 2. Deep pedestrian classification based on the matting transparency mask

13:   Pedestrian classification using AlexNet with the alpha matte obtained by preprocessing

1.4.2.2 Preprocessing algorithm for infrared pedestrian images based on automatic image matting

The preprocessing algorithm based on automatic image matting is the key to the infrared pedestrian classification algorithm enhanced by automatic image matting. Its essence is an automatic image matting algorithm for infrared pedestrian images, which is used to solve the problem of difficult contour feature extraction when the brightness of the pedestrian in the far-infrared image is similar to the background. The preprocessing algorithm mainly includes two steps: automatic generation of the pedestrian trimap in the infrared image and far-infrared image matting. The former automatically generates the corresponding pedestrian trimap for the ROI in the far-infrared image according to the estimated position of the pedestrian's head and upper body. The latter clips the far-infrared image to output an alpha matte to realize precise extraction of the foreground region.

1.4.2.2.1 Automatic algorithm for generating pedestrian trimap in infrared images

The image matting technology can accurately extract the foreground from a cluttered background, the image matting technology is expected to solve the interference problem caused by cluttered backgrounds in pedestrian classification based on infrared images. However, because natural image matting algorithms require a trimap as input, existing methods for generating trimaps rely on manual labeling and are unable to generate trimaps for natural images. These limits the application of image matting in infrared pedestrian classification.

The trimap is an important input in natural image matting, and the quality of the trimap directly affects the quality of the obtained alpha matte. Natural image matting estimates the alpha value of unknown region pixels based on the similarity between unknown region pixels and known region pixels. The known and unknown regions are precisely divided by the trimap.

The traditional method of generating a trimap involves manually marking the regions of interest, which is not feasible for a large number of images. Existing automatic trimap generation methods rely on special shooting techniques such as flash photography and fixed background shooting. These automatic trimap generation methods cannot generate trimaps for a large number of images shot in natural conditions. Therefore, existing trimap generation methods are not suitable for automatic pedestrian classification tasks. To address this problem, this section introduces an algorithm for automatically generating trimaps of pedestrian infrared images. The algorithm does not rely on user interaction but generates a trimap based on prior knowledge of pedestrian classification in infrared images. The generated trimap provides the head and upper body areas of the pedestrian and some background areas, and the rest is an unknown area. The generated trimap provides the necessary information for far-infrared image matting, which is the key to achieving pedestrian preprocessing of infrared images based on fully automatic matting. The algorithm for automatically generating trimaps of pedestrian infrared images assumes that the ROI in the input far-infrared image contains walking pedestrians and tries to generate a suitable trimap for pedestrians. As shown in Fig. 1.22, the algorithm mainly consists of three steps: pedestrian head positioning, pedestrian upper body positioning, and trimap generation.

In order to adapt to changes in the posture of pedestrians, the algorithm for automatically generating trimaps of infrared image pedestrians for locating key parts of the pedestrian's body refers to anatomical research [49] and professional knowledge in the field of infrared image pedestrian classification, and the pedestrian key body part location strategy is designed based on human body part model in infrared images. The localization of each part of the pedestrian is a key link in the automatic generation of trimaps of pedestrians. However, the appearance and posture of the pedestrian in the ROI are constantly changing, which presents a challenge for pedestrian localization. Once there is a deviation in the localization of the body parts of the pedestrian, the resulting trimaps will contain incorrectly labeled areas, resulting in a significant decrease in the quality of foreground extraction. Inspired by the work of Lee et al. [49], the algorithm in the infrared image pedestrian trimap automatic generation method uses a human body part model to fully utilize the constraints between each part. Due to interference from factors such as cluttered backgrounds, changes in pedestrian posture, and changes in observation angles, it is very difficult to accurately locate all parts of the human body in the ROI for far-infrared images. Considering

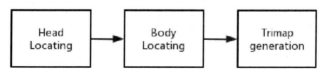

Figure 1.22 Schematic diagram of the main steps of the automatic generation algorithm for infrared image pedestrian trilateration.

that foreground extraction can be achieved by labeling the foreground area in natural image matting, it is not necessary to locate all parts of the human body. Therefore, the infrared image pedestrian trimap automatic generation algorithm only locates two key parts. That is, the pedestrian's head and trunk that can be robustly located.

Since the far-infrared image is based on thermal imaging, the brightness of the head area of the human body is usually higher than that of the background. The location of the head area in the far-infrared image can make use of additional prior knowledge. The brightness of the human body's upper body may not be higher than that of the background area due to the influence of clothing. Therefore, the location of the head of the human body has more prior knowledge, and the positioning result is more accurate and robust. This algorithm for generating a three-part image of an infrared image pedestrian first locates the head considering these factors and then uses constraints between the head and upper body to locate the upper body. The introduced algorithm first standardizes the ROI in the far-infrared image before locating the head, specifically by scaling the input image to $w \times h$ size, where w and h respectively represent the width and height of the standardized image. The size of the standardized image in this algorithm is the same as that of most pedestrian datasets in far-infrared images, set to 32×64. The methods for locating the head and upper body of a pedestrian are introduced separately below.

As shown in Fig. 1.23, pedestrian head positioning includes two steps: rough positioning and precise positioning of the head.

The goal of coarse localization of the head is to find the approximate position of the pedestrian's head, specifically by finding a pixel belonging to the head region. The coarse localization approach of the head is designed based on the prior knowledge that the head region has higher brightness. Firstly, the Otsu algorithm is used to binarize the normalized image. Then the pixels with a true value are selected to form a candidate pixel set. The cluttered background area may contain noise, which causes inaccurate positioning. Through observation, we found that the pedestrian's head is usually in the middle position of the upper part of the

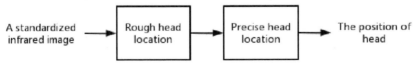

Figure 1.23 Pedestrian head positioning flowchart.

image, while noise often appears on both sides. Therefore, we added spatial position constraints to the coarse localization of the head. The head coarse localization can be modeled as an optimization problem.

$$\min_{(x,y)\in\Omega} D\left(x_0^h, y_0^h, x, y, \lambda\right) \quad s.t. \ f_B(x, y) = 1 \tag{1.30}$$

$f_B(x,y)$ represents the value of the pixel at coordinate (x,y) in a binary image, and $D(x_0^h, y_0^h, x, y, \lambda)$ represents the distance between the initial position (x_0^h, y_0^h) and (x, y) of the human head and (x_0^h, y_0^h). The initial position of the human head is $(\lfloor w/2\rfloor, \lfloor h/6\rfloor)$ Its definition is as follows:

$$D\left(x_0^h, y_0^h, x, y, \lambda\right) = \sqrt{\left(x_0^h - x\right)^2 + \mu\left(y_0^h - y\right)^2} \tag{1.31}$$

μ is the weight that balances the cost of horizontal distance and vertical distance. When $\mu = 1, D(x_0^h, y_0^h, x, y, \lambda)$ represents the Euclidean distance between (x_0^h, y_0^h) and (x, y). Considering that the human head is usually in the middle position, the cost of horizontal distance should be higher than that of vertical distance, so the value of λ in Eq. (1.30) should be greater than 1. The optimization problem of coarse location of the human head can be quickly solved by traversing the binary image in the order of increasing distance of $D(x_0^h, y_0^h, x, y, \lambda)$ and finding the first pixel that takes true.

The goal of precise head positioning is to locate the center point of a pedestrian's head. Because the human body's temperature is generally higher than that of the background in most cases, the human body's head area radiates more energy than the background area, human body is represented as brighter in infrared images. In addition, when imaging in the infrared spectrum, the central area of the head has more energy that can enter the sensor, making the center area brighter than the edge area. Therefore, the center point of a pedestrian's head in an infrared image is a local maximum. Precise head positioning of a pedestrian takes the results of coarse head positioning $(\tilde{x}^h, \tilde{y}^h)$ as input and searches for the center position of the pedestrian's head in a local area. At last, the location of the pixel with the highest brightness is used as the center position of the pedestrian's head. Precise head positioning of a pedestrian can be modeled as the following optimization problem.

$$\max_{(x,y)\in\Omega} f_S(x, y) \quad s.t. \ D\left(\tilde{x}^h, \tilde{y}^h, x, y, 1\right) \le h_1 \tag{1.32}$$

The algorithm for generating the three-part infrared image of a pedestrian uses the hill climbing method to achieve precise positioning of the head. The infrared image is standardized, and $f_S(x,y)$ represents the value of the pixel at position (x,y) in the image, while Ω represents the set of pixels in the standardized infrared image. The process of determining the center of the head within a local area centered on a given rough location can be considered as a single peak search, as the brightness of the center of the head in the infrared image is a local maximum. Considering the stability and speed of the search, the hill climbing method is selected to achieve precise positioning of the head. Algorithm 1.6 provides the pseudocode for the hill climbing method used, where w_1 and h_1 represent the width and height of the bounding box for the head region, respectively, and their values are $\lfloor w/10 \rfloor$ and $\lfloor h/16 \rfloor$.

As mentioned above, in far-infrared images, the human body is relatively brighter due to its body temperature, human body is easily affected by clothing. The upper body localization is achieved by estimating the offset between the head and upper body, based on the assumption that the human upper body is positioned upright while walking, by fully utilizing the relatively robust head localization information. Inspired by the study by Lee et al., the upper body ROI is first defined based on the head localization results. The cumulative histogram of the upper body ROI is obtained by vertically projecting each pixel in the ROI.

Fig. 1.24 shows the position of the upper body ROI with respect to the center of the head localization results. The black dots above indicate the head localization results, and the gray area below indicates the upper

---

ALGORITHM 1.6 Hill climbing algorithm for precise head location

input: $\tilde{x}^h, \tilde{y}^h, h_1$.

output: x^h, y^h

1: $x^h \leftarrow \tilde{x}^h$, $y^h \leftarrow \tilde{y}^h$, MaxIterations $\leftarrow h_1$, $i \leftarrow 0$

2: **while** $i <$ MaxIterations **do**

3:    $xt \leftarrow xh$, $yt \leftarrow yh$

4:       **for each** $m \in \{-1,0,1\}$ **do**

5:          **for each** $n \in \{-1,0,1\}$ **do**

6:             **if** $f_S(x^h + m, y^h + n) > f_S(x^t, y^t)$ **then**

7:                $x^t \leftarrow x^h + m$, $y^t \leftarrow y^h + m$

8:             **end if**

9:          **end for**

10:    **end for**

11:    $xh \leftarrow xt$, $yh \leftarrow yt$

12: **end while**

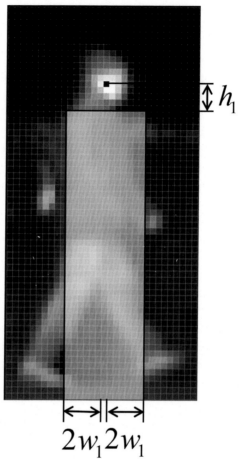

$2w_1 2w_1$

Figure 1.24 Schematic diagram of the region of interest of the human upper body.

body ROI. Existing research usually takes the x-coordinate corresponding to the maximum value in the accumulation histogram as the horizontal offset of the upper body. In contrast, the hierarchical matting algorithm takes the mean value of the x-coordinates corresponding to the top 50% of values in the accumulation histogram as the horizontal offset x^o. By weighting multiple positions with strong energy in the cumulative histogram, the impact of noisy backgrounds is suppressed. The center position of the human upper body can be obtained by Eq. (1.33):

$$\left(x^b, y^b\right) = \left(x^h + x^o, y^h + h_2/2\right) \tag{1.33}$$

where h_2 represents the height of the human upper body, which is equal to $\lfloor 3h/8 \rfloor$. In addition, the localization of the human upper body is constrained based on research on the human body model [49], the constraint is that the angle between the line connecting the center position of the head and the center position of the upper body and the perpendicular line is no more than $12°$. Once this constraint is not satisfied, the horizontal offset x^o of the human upper body will be set to zero, and the center position of the upper body will be recalculated.

In the trimap generation method, the trimap of pedestrians is generated based on the trimap template of the center of the head and trunk of the human body obtained from the above steps. The initial trimap is set as a single-channel image with 0 values of the same size as the standardized image. The anchor points of the head trimap template and the anchor points of the trunk trimap template are aligned with the estimated center position of the human head and the center position of the human body trunk, respectively, while the anchor points of the human lower limb trimap template are aligned with the estimated center point of the trunk. The trimap templates for the head, trunk, and lower limbs are shown in Fig. 1.25, where the anchor points in the template are represented by asterisks. When the template areas overlap, the larger value is retained. For example, when the white and gray areas overlap, the white is retained. When the gray and black areas overlap, the gray is retained.

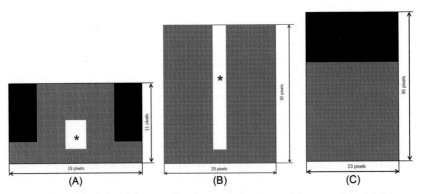

Figure 1.25 Template of human head, upper body, and lower extremity trimaps used in trilateration generation. (A) Head template. (B) Upper body template. (C) Lower body template.

1.4.2.2.2 Far-infrared pedestrian extraction

The research goal of far-infrared image matting is to use far-infrared images and their corresponding trimap to generate an alpha matte that provides clear contour features. The far-infrared image matting enhances the pedestrian area and suppresses irrelevant cluttered background areas effectively, providing a clear pedestrian contour for pedestrian classification.

Due to the better robustness of sampling-based matting algorithms in the presence of interference, sampling-based matting algorithms are more suitable for the matting task of enhancing human images in far-infrared images, compared with propagation-based matting algorithms.

As shown in Fig. 1.26, Fig. 1.26A is an infrared image; Fig. 1.26B is a trimap; Fig. 1.26C is the alpha matte obtained by propagation-based

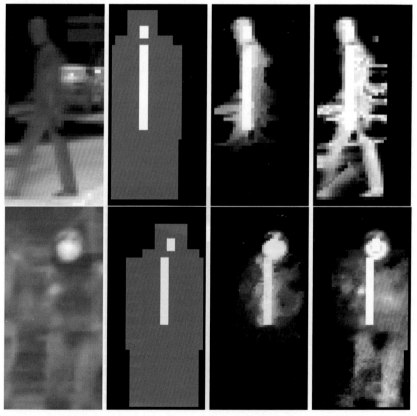

Figure 1.26 Comparison of alpha mattes obtained by sampling-based and propagation-based image matting algorithms in the presence of interference. (A) Infrared image. (B) Trimap. (C) The propagation-based image matting algorithm named KNN matting [13]. (D) The sampling-based image matting algorithm named global matting [3]. *KNN*, K-nearest neighbors.

matting algorithm (K-nearest neighbor matting algorithm); Fig. 1.26D is the alpha matte obtained by sampling-based matting algorithm (sampling-based global matting algorithm). The propagation-based matting algorithm is easily disturbed by noise during the propagation of alpha value, which leads to the poor extraction of the pedestrian's legs; the sampling-based matting algorithm can still provide high-quality alpha mattes in the presence of noise interference; the machine learning-based matting algorithm is not suitable for this task due to the lack of training data.

In far-infrared image matting, we chose He's global sampling-based segmentation algorithm after balancing between segmentation speed and accuracy [3]. The algorithm includes three steps: known pixel sampling, foreground/background pixel pair evaluation, and alpha value estimation.

In known pixel sampling, only pixels that fall on the edge of the known area are sampled as samples. For color visible light images, the optimal foreground/background pixel pairs are often not selected as samples because they do not fall on the edge of the known area, resulting in the problem of losing the optimal foreground/background pixel pairs. However, this problem is not easy to occur in far-infrared images. Because the resolution of far-infrared images is much lower than that of color visible light images due to the different imaging principles, a larger proportion of known pixels will fall in the edge area of the known area. The image named GT01 in the segmentation benchmark dataset introduced by Rhemann et al. [50] is used as an example for explanation. The resolution of this color image is 800×497, and the known foreground area of the three-part image contains 2372 pixels, 1.1% of which fall on the edge of the foreground area. If the image and three-part image are scaled to the same size as the standardized image of 32×64, the known foreground area of the scaled three-part image contains 172 pixels, 15.7% of which fall on the edge of the known foreground area. Due to the significantly increased proportion of pixels covered by the edge of the known area, the problem of losing the optimal foreground/background pixel pairs can be effectively alleviated in far-infrared images.

The Cartesian product of the foreground sample set and the background sample set obtained by sampling can generate a set of foreground/background pixel candidate sets Γ. Each candidate pixel pair is evaluated by the foreground/background pixel pair evaluation function to select the optimal foreground/background pixel pairs. Since infrared images cannot provide color or texture information, the foreground/background pixel pair evaluation function designed for color visible light images cannot be directly applied to infrared image matting. Therefore, in the infrared image matting, we introduce a foreground/ background pixel pair evaluation function for infrared image matting based on the objective function of the global sampling-based segmentation algorithm.

$$\varepsilon = \varepsilon_i + \varepsilon_s \tag{1.34}$$

where ε_i and ε_s represent brightness distortion and spatial cost, respectively. Their definitions are as follows:

$$\varepsilon_i = \left| i^{(U)} - \left(\hat{\alpha} i_k^{(F)} + (1 - \hat{\alpha}) i_k^{(B)} \right) \right| \tag{1.35}$$

$$\varepsilon_S = \frac{\left\| S^{(U)} - S_k^{(F)} \right\|}{\min\limits_{j=1}^{|\Gamma|} \left\{ \left\| S^{(U)} - S_j^{(F)} \right\| \right\}} + \frac{\left\| S^{(U)} - S_k^{(B)} \right\|}{\min\limits_{j=1}^{|\Gamma|} \left\{ \left\| S^{(U)} - S_j^{(B)} \right\| \right\}} \tag{1.36}$$

where $i^{(U)}$ and $s^{(U)}$ represent the brightness and spatial coordinates of the given unknown pixel respectively. $i_k^{(F)}$ and $i_k^{(B)}$ represent the brightness of the foreground and background pixels of the kth candidate foreground/ background pixel pair in the set Γ. $s_k^{(F)}$ and $s_k^{(B)}$ represent the spatial coordinates of the foreground and background pixels of the kth candidate foreground/background pixel pair in the set Γ. $\hat{\alpha}$ represents the estimated alpha value of the given unknown pixel, which is calculated as follows:

$$\hat{\alpha} = \min \left\{ 1, \frac{\left| i_z^{(U)} - i_k^{(B)} \right|}{\left| i_k^{(F)} - i_k^{(B)} \right| + \varepsilon} \right\} \tag{1.37}$$

where ε is a small constant used to avoid division by zero errors. Once the optimal foreground/background pixel pairs are determined, the alpha matte for the segmentation can be calculated using Eq. (1.37).

1.4.2.2.3 Pedestrian classification algorithm based on depth infrared images using alpha mattes for segmentation

A pedestrian classification algorithm for deep infrared images based on the alpha matte is introduced. It utilizes a deep neural network to learn representations of pedestrians from the alpha matte of the far-infrared image obtained through automated segmentation. The reason for choosing a deep neural network is twofold. Although the contour of pedestrians is enhanced in the alpha matte of the far-infrared image obtained through automated segmentation, there is a lack of the feature extraction and classification algorithm specifically designed for the alpha matte in research on expert-driven algorithms. Data-driven algorithms, represented by deep neural networks, have proven to be capable of automatically learning structural features that are difficult for humans to discover from high-dimensional data for classification. Existing algorithms usually use very deep neural networks to handle complex classification scenarios. However, the higher computational cost limits the application of very deep neural networks in practical scenarios. Therefore, it is necessary to explore deep infrared pedestrian classifiers with accurate classification achieved through prior knowledge of infrared pedestrian classification rather than deepening the neural network.

The depth infrared image pedestrian classification algorithm based on the alpha matte is introduced in this section, and it works on the basis of hypothesis verification. The algorithm assumes that each input infrared image contains pedestrians, and the preprocessing of infrared image pedestrians based on automatic image matting is performed on each input image. The working principle of this algorithm is as follows: when the input image indeed contains pedestrians, clear pedestrian contours will appear in the alpha matte generated by the preprocessing; when the input image does not contain pedestrians, the foreground region known in the generated three-part chart corresponds to nonpedestrian objects, and the contours of the foreground objects in the alpha matte obtained by the image matting do not match those of pedestrians. The deep learning classification algorithm can effectively distinguish between pedestrian targets and nonpedestrian targets by using the alpha matte with clear contour features.

Considering that high computational cost would limit the industrial application of the introduced infrared pedestrian classification algorithm, this section compares several representative representation learning algorithms based on deep neural networks. As shown in Table 1.12, increasing

Table 1.12 Three existing advanced deep neural network key performance indicators.

Deep neural network	Layer number	Time consumed by model prediction (s)	Model size (MB)
AlexNet [45]	8	0.243	227.5
VGG [46]	16	1.435	537.1
ResNet [47]	152	1.153	241.4

the number of network layers often leads to an increase in computation time and storage space. The infrared pedestrian classification algorithm based on pedestrian matting selected AlexNet [45] as the pedestrian representation learning algorithm after balancing between classification performance and computational cost, and used pre-training models to improve its training accuracy.

1.4.2.3 Experimental results and discussion

In this section, the effectiveness of the pedestrian classification algorithm based on automatic image matting enhancement is verified through five experiments. The experiments were conducted on a computer equipped with a 3.1 GHz Intel Core i5 processor and 8 GB of memory. The deep neural network was implemented on the Cafe platform [51]. The automatic image matting algorithm was implemented using $C + +$ language.

The experiment was conducted on three large infrared image datasets: the LSI Infrared Pedestrian Dataset [52], the RIFIR Infrared Dataset [53], and the KAIST Multispectral Pedestrian Detection Benchmark Dataset [54]. The LSI dataset collected 81,592 infrared images using a vehicle-mounted camera, each corresponding to a ROI for pedestrian detection, with an image size of 32×64. Among them, 53,598 images were used as the training set, and 27,994 images as the test set. The training set contained 10,208 positive samples and 43,390 negative samples, while the test set contained 5,944 positive samples and 22,050 negative samples. The RIFIR dataset contains 24,395 pairs of infrared and visible light images captured by a vehicle-mounted camera, with an image size of 48×96. The training set of the RIFIR dataset contains 34,810 images, including 9,202 positive samples and 25,608 negative samples. The test set contains 26,477 images, including 2,034 positive samples and 24,443 negative samples. The KAIST dataset marks 30,970 pedestrian areas (i.e., positive samples) in 95,324 pairs of infrared and visible light images. Among them, 12,856 were used for training and 18,114 were used for testing, with the sizes of the marked pedestrian areas varying. Since negative sample labels were not provided in the dataset, two images were randomly cropped with a size of 32×64 in a far-infrared image that does not include pedestrians in the experiment, generating 43,626 negative samples. Only infrared images were used in the experiment, and visible light images were not involved.

In the experiment, precision [55], recall [56], accuracy [57,58], and F-value(F-measure) [59] are four commonly used evaluation metrics in machine learning to quantitatively evaluate the performance of different methods in the pedestrian classification task in infrared images. The larger the value of the evaluation metric, the better the performance of the classifier.

The infrared image pedestrian classification algorithm based on automatic matting enhancement uses AlexNet to train based on the stochastic gradient descent method recommended in reference [45]. The momentum is set to 0.9, and the weight decay is set to 0.0005. The batch size is set to 32 during training. Firstly, the λ parameter involved in the generation of the trimap in the infrared pedestrian classification algorithm based on automatic image matting enhancement is discussed through experiments. Table 1.13 summarizes the performance indicators of the introduced algorithm on the LSI dataset under different values of λ. It can be found from the table that the value of λ has little influence on the classification performance of this algorithm, indicating that the algorithm is not sensitive to the setting of the λ parameter. Given that the algorithm achieves the best performance in three of the four evaluation indicators when $\lambda = 3$, $\lambda = 3$ is chosen for the following experiments.

1.4.2.3.1 Classification performance verification experiment of pedestrian classification algorithm based on automatic image matting enhancement of infrared images

The purpose of the first experiment is to verify the performance of the infrared image pedestrian classification algorithm based on automatic image matting enhancement in infrared image pedestrian classification tasks.

This experiment used five existing infrared image pedestrian classification algorithms as performance benchmarks, two of which are preprocessing-based infrared image pedestrian classification algorithms IPS-AlexNet and Minmax-ACF [60], and the other three are advanced deep neural network-based classification algorithms AlexNet [45], VGG [46], and ResNet [47]. Among them, IPS-AlexNet is an infrared image pedestrian classification algorithm generated by using the automatic infrared image pedestrian

Table 1.13 Classification performance of infrared image pedestrian classification algorithm based on automatic pedestrian mattings on LSI dataset with different values of λ. Quantitative comparison of classification performance.

The value of λ	$\lambda = 1$	$\lambda = 3$	$\lambda = 5$
Precision	99.49%	**99.68%**	99.64%
Recall	**99.14%**	99.09%	99.04%
Accuracy	99.32%	**99.39%**	99.34%
F-measure	99.32%	**99.38%**	99.34%

Note: The best results are highlighted in bold.

segmentation preprocessing results introduced in the literature [61] as the input of the AlexNet classifier. The thresholds of each classifier were obtained by selecting the point that maximizes the area under the receiver operating characteristic (ROC) curve.

Table 1.14 summarizes the quantitative comparison results of the infrared pedestrian classification algorithm based on automatic image matting enhancement with five existing infrared pedestrian classification algorithms on LSI, RIFIR, and KAIST datasets. It can be seen from the table that the introduced algorithm has achieved better classification performance on almost all performance indicators than other classification algorithms on these three datasets, which indicates that this algorithm has better classification performance in infrared pedestrian classification tasks. The experiment further compares the average performance indicators of the infrared pedestrian classification algorithm. The average performance indicators are obtained by repeating the experiment five times and calculating the mean and standard deviation of each performance indicator (as shown in Table 1.14). Compared with ResNet, this algorithm has higher means and smaller standard deviations for four performance indicators, indicating that the algorithm not only achieves better performance in infrared pedestrian classification but also has more stable training results than ResNet. This experimental result further confirms the effectiveness of the infrared pedestrian classification algorithm based on automatic image matting enhancement in infrared pedestrian classification tasks.

1.4.2.3.2 Analysis of the performance improvement of pedestrian classification based on infrared image preprocessing using automatic matting

The second experiment is conducted to discuss the reasons for the performance improvement of the infrared pedestrian classification algorithm. The difference lies in the fact that the classic AlexNet uses raw far-infrared images as input, while the introduced algorithm uses the alpha mattes generated from infrared pedestrian images processed by automatic image matting as the input of the AlexNet. Therefore, the performance improvement is brought by the alpha matte generated from automatic image matting of infrared pedestrian images.

Considering that the brightness of pedestrians in far-infrared images varies with weather and seasons, which poses a significant challenge for pedestrian classification in infrared images, three types of data were used in the experiment: the first type is positive sample images where the entire

Table 1.14 Quantitative comparison of the classification performance of the infrared image pedestrian classification algorithm based on automatic pedestrian matting with five existing infrared image pedestrian classification algorithms on the LSI, RIFIR, and KAIST datasets.

Metrics (dataset)	Infrared pedestrian classification algorithm	IPS-AlexNet [61]	Minmax-ACF [60]	AlexNet [45]	VGG [46]	ResNet [47]
Precision (LSI)	**99.61%** (99.68 ± 0.13%)	92.49%	99.50%	97.78%	95.76%	97.96% (99.21 ± 0.76%)
Recall (LSI)	99.33% (99.06 ± 0.18%)	81.63%	74.16%	99.26%	99.14%	**99.38%** (98.65 ± 0.77%)
Accuracy (LSI)	**99.77%** (99.43 ± 0.19%)	87.50%	94.43%	99.36%	98.89%	99.43% (99.09 ± 0.37%)
F-measure (LSI)	**99.47%** (99.37 ± 0.07%)	86.72%	84.98%	98.51%	97.42%	98.66% (98.93 ± 0.35%)
Precision (RIFIR)	97.05%	87.93%	76.47%	90.51%	**97.41%**	94.73%
Recall (RIFIR)	**98.72%**	90.22%	84.66%	92.82%	94.40%	96.41%
Accuracy (RIFIR)	**97.86%**	88.92%	96.82%	91.55%	95.94%	95.53%
F-measure (RIFIR)	**97.88%**	89.06%	80.35%	91.65%	95.88%	95.57%
Precision (KAIST)	**98.36%**	86.33%	95.16%	94.29%	93.06%	93.47%
Recall (KAIST)	**95.05%**	77.82%	73.78%	94.32%	92.99%	94.98%
Accuracy (KAIST)	**96.73%**	82.75%	87.33%	94.31%	93.03%	94.17%
F-measure (KAIST)	**96.68%**	81.85%	83.11%	94.31%	93.02%	94.22%

Note: 1. Infrared pedestrian classification algorithm and the one closest in performance to it, ResNet, were run five times on the LSI dataset to obtain the mean and standard deviation of performance indicators, expressed as (mean ± standard deviation). The best results are highlighted in bold.

pedestrian area is brighter than the background; the second type is positive sample images where part of the pedestrian area is darker than the background; and the third type is negative sample images that do not contain pedestrians. Fig. 1.27 shows the comparison between the input image

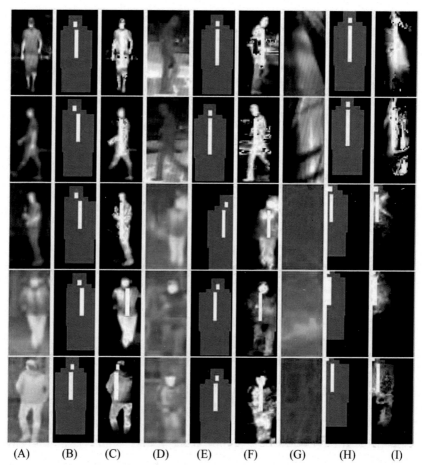

(A)   (B)   (C)   (D)   (E)   (F)   (G)   (H)   (I)

Figure 1.27 Comparison image data of AlexNet input image (i.e., alpha matte) and classical AlexNet input image (i.e., original far-infrared image region of interest) in the automatic image matting enhancement-based pedestrian classification algorithm for infrared images from LSI and RIFIR datasets. Examples are from LSI [52] and RIFIR [53] far-infrared pedestrian dataset. (A) Positive infrared images in which all of pedestrian regions appear brighter than nonpedestrian regions. (D) Positive infrared images in which part of pedestrian regions appear darker than nonpedestrian regions. (G) Negative infrared images. (B), (E), and (H) automatically generated trimaps with respect to (A), (D), and (G), respectively. (C), (F), and (I) Alpha mattes regarding (A), (D) and (G), respectively.

(i.e., the alpha matte) of AlexNet in the infrared image pedestrian classification algorithm based on automatic image matting enhancement and the classic AlexNet input image (i.e., the ROI in the original far-infrared image). Fig. 1.27A shows a positive sample example where the pedestrian area is brighter than the background area; Fig. 1.27D shows a positive sample example where part of the pedestrian area is darker than the background area; Fig. 1.27G shows a negative sample example; Fig. 1.27B, E, and H show the trisection graph automatically generated by the infrared image pedestrian trimap generation algorithm on the samples in Fig. 1.27A, D, and G respectively; Fig. 1.27C, F, and I show the image matting alpha mattes calculated by the infrared image pedestrian preprocessing algorithm based on automatic image matting enhancement on the samples in Fig. 1.27A, D, and G, respectively. It can be seen from Fig. 1.27C and F that the contour of the pedestrian is significantly enhanced and the cluttered background is effectively suppressed after the infrared image pedestrian preprocessing based on automatic image matting enhancement. Even in the samples where part of the pedestrian area is darker than the background area, the infrared image pedestrian preprocessing algorithm can output consistent enhanced pedestrians and suppressed backgrounds. In the negative sample, as shown in Fig. 1.27I, the clear contour features presented in the alpha matte obtained by the preprocessing algorithm are obviously different from those of pedestrians. This is because the foreground area known in the generated trisection graph corresponds to nonpedestrian objects, so the foreground extracted in the alpha matte is nonpedestrian objects, and their contour is different from that of pedestrians. Since the infrared image pedestrian preprocessing algorithm can output image matting alpha mattes with significant contour features for positive and negative samples in various situations, the infrared pedestrian classification algorithm can accurately distinguish pedestrian and nonpedestrian targets from the preprocessing results, while classical classification algorithms based on deep neural networks may cause classification errors due to changes in pedestrian brightness.

1.4.2.3.3 Analysis of the impact of infrared image pedestrian preprocessing based on automatic image matting on the performance of deep learning classification algorithms

The purpose of the third experiment is to verify whether the infrared image pedestrian preprocessing based on automatic image matting can help improve the classification performance of deep neural networks for infrared image pedestrians. This experiment involves two types of infrared image pedestrian classification algorithms: a deep neural network classification algorithm based on automatic image matting of infrared image pedestrians and a classic deep neural network classification algorithm. The former uses the alpha matte obtained by the fully automatic infrared pedestrian preprocessing as described in Section 1.4.2 as the input image in training and testing, while the latter uses the original version of the far-infrared image ROI as the input image This experiment uses the ROC curve as a performance indicator. The ROC curve describes the true positive rate and false positive rate of the classifier at different thresholds. The higher the true positive rate and the lower the false positive rate, the better the classification performance.

Figs. 1.28−1.30 show the ROC curve comparison of a deep neural network classification algorithm based on automatic infrared image human

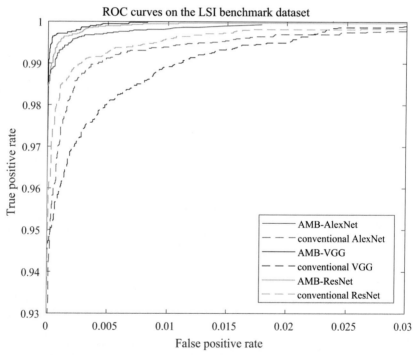

Figure 1.28 Comparison of ROC curves obtained on LSI dataset by deep neural network classification algorithm based on automatic image matting for pedestrian preprocessing of infrared images and classical neural network classification algorithm. *ROC,* Receiver operating characteristic.

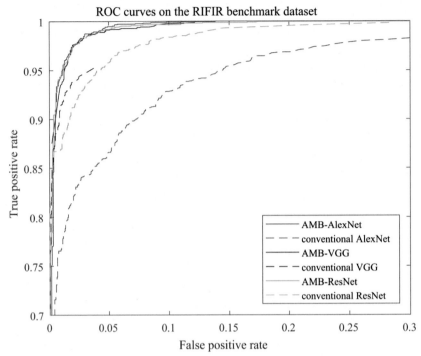

ROC curves on the RIFIR benchmark dataset

Figure 1.29 Comparison of ROC curves obtained on RIFIR dataset by deep neural network classification algorithm based on automatic image matting for infrared image pedestrian preprocessing and classical neural network classification algorithm. *ROC*, Receiver operating characteristic.

preprocessing using background abstraction and classical deep neural network classification algorithm on three datasets, respectively. The solid line represents the deep neural network classification algorithm based on automatic infrared image human preprocessing using background abstraction, while the dotted line represents the classical deep neural network classification algorithm. The same color indicates that the deep neural network structure used is the same. The same color representation uses the same deep neural network structure.

The deep neural network classification algorithm based on automatic image matting of infrared images for pedestrian preprocessing has AUC indicators that are superior to classical methods using the same deep neural network on both the LSI and RIFIR datasets. Moreover, at low false positive rates, the true positive rate of the deep neural network classification algorithm based on automatic image matting of infrared images for pedestrian preprocessing is significantly higher than that of the corresponding

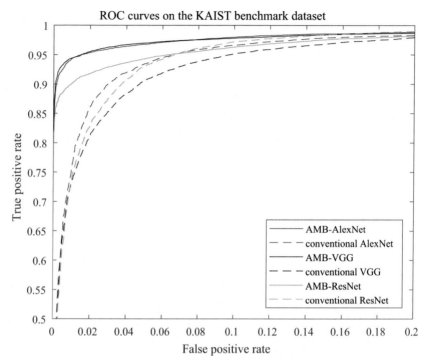

Figure 1.30 Comparison of ROC curves obtained on KAIST dataset by deep neural network classification algorithm based on automatic image matting for pedestrian preprocessing of infrared images and classical neural network classification algorithm. *ROC*, Receiver operating characteristic.

classical deep neural network classification algorithm. This experimental result is attributed to the introduced automatic image matting of infrared images for pedestrian preprocessing that enhances the contour features of the foreground target, suppresses cluttered backgrounds, and provides consistent normalized enhancement for pedestrian targets in different far-infrared images, infrared images for pedestrian preprocessing helps distinguish between pedestrians and nonpedestrians. Therefore, the deep neural network classification algorithm based on automatic image matting of infrared images for pedestrian preprocessing achieves better classification performance. This result further demonstrates that the automatic image matting of infrared images for the pedestrian preprocessing algorithm can effectively improve the performance of deep neural networks. It is worth noting that in the KAIST dataset experiment, the true positive rate of the ResNet based on automatic image matting of infrared images for pedestrian preprocessing is higher than that of the classical ResNet at a false positive rate below 0.068, but lower than that of the classical ResNet when the false positive rate is above 0.068. The ResNet with more layers based on automatic image matting of infrared images for pedestrian preprocessing has lower classification performance on the KAIST dataset than the VGG and AlexNet based on automatic image matting of infrared images for pedestrian preprocessing, but higher performance than the other two deep neural network classification algorithms based on automatic image matting of infrared images for pedestrian preprocessing on the LSI and RIFIR datasets. Since all three algorithms based on automatic image matting of infrared images for pedestrian preprocessing use the same data for training and testing, this phenomenon is attributed to the complexity of the KAIST dataset and the difficulty of training with a ResNet containing 152 neural layers on complex datasets. On the one hand, the positive samples in the KAIST dataset are small and have various sizes, making it difficult for the deep neural network to learn the representation information of infrared image pedestrians. On the other hand, training a ResNet that includes 152 neural layers on a complex dataset is prone to getting trapped in local optimal solutions.

To quantify the performance differences between the deep neural network classification algorithms based on automatic image matting of infrared pedestrian preprocessing and classical deep neural network classification algorithms, this experiment further uses the four evaluation indicators of accuracy, recall, precision, and F-value to perform a quantitative comparison of the classification performance of the two types of algorithms on three datasets. The threshold of each classifier is obtained by selecting the point that maximizes the area below the ROC curve. The threshold setting of the classical deep neural network classification algorithm and AlexNet based on the automatic image matting of infrared pedestrian preprocessing is consistent with the first experiment. The threshold settings of VGG based on automatic image matting of infrared pedestrian preprocessing in LSI, RIFIR, and KAIST are 0.00722, 0.96278, and 0.28288, respectively. The threshold settings of ResNet based on automatic image matting of infrared pedestrian preprocessing in LSI, RIFIR, and KAIST are 0.46972, 0.62296, and 0.30958, respectively. The experimental results are shown in Tables 1.15–1.17.

As shown in these tables, deep neural network classification algorithms based on automatic image matting of infrared pedestrian preprocessing outperform corresponding classic algorithms in almost all performance

Table 1.15 Comparison of quantitative infrared image pedestrian classification performance of deep neural network classification algorithms based on automatic image matting for infrared image pedestrian preprocessing with classical deep neural network classification algorithms on LSI dataset.

Infrared pedestrian classification algorithm	Precision	Recall	Accuracy	F-measure
AlexNet based on infrared image preprocessing using automatic pedestrian matting	**99.61%**	99.33%	**99.77%**	**99.47%**
VGG based on infrared image preprocessing using automatic pedestrian matting	98.89%	99.68%	99.69%	99.28%
ResNet based on infrared image preprocessing using automatic pedestrian matting	98.85%	**99.83%**	99.72%	99.34%
AlexNet [45]	97.78%	99.26%	99.36%	98.51%
VGG [46]	95.76%	99.14%	98.89%	97.42%
ResNet [47]	97.96%	99.38%	99.43%	98.66%

The best results are highlighted in bold.

Table 1.16 Comparison of quantitative infrared image pedestrian classification performance of deep neural network classification algorithms based on automatic image matting for infrared image pedestrian preprocessing with classical deep neural network classification algorithms on RIFIR dataset.

Infrared pedestrian classification algorithm	Precision	Recall	Accuracy	F-measure
AlexNet based on infrared image preprocessing using automatic pedestrian matting	97.05%	**98.72%**	**97.86%**	**97.88%**
VGG based on infrared image preprocessing using automatic pedestrian matting	**98.02%**	97.59%	97.81%	97.81%
ResNet based on infrared image preprocessing using automatic pedestrian matting	97.79%	97.74%	97.76%	97.76%
AlexNet [45]	90.51%	92.82%	91.55%	91.65%
VGG [46]	97.41%	94.40%	95.94%	95.88%
ResNet [47]	94.73%	96.41%	95.53%	95.57%

The best results are highlighted in bold.

Table 1.17 Comparison of quantitative infrared image pedestrian classification performance of deep neural network classification algorithms based on automatic image matting for infrared image pedestrian preprocessing with classical deep neural network classification algorithms on KAIST dataset.

Infrared pedestrian classification algorithm	Precision	Recall	Accuracy	F-measure
AlexNet based on infrared image preprocessing using automatic pedestrian matting	**98.36%**	95.05%	96.73%	96.68%
VGG based on infrared image preprocessing using automatic pedestrian matting	98.11%	**95.40%**	**96.78%**	**96.74%**
ResNet based on infrared image preprocessing using automatic pedestrian matting	96.72%	93.00%	94.92%	94.82%
AlexNet [45]	94.29%	94.32%	94.31%	94.31%
VGG [46]	93.06%	92.99%	93.03%	93.02%
ResNet [47]	94.29%	94.32%	94.31%	94.31%

The best results are highlighted in bold.

indicators. The only exception is ResNet based on automatic image matting of infrared pedestrian preprocessing, which has lower recall indicators than classic deep neural network classification algorithms on KAIST dataset.

The ROC curve and quantitative comparison analysis results both indicate that the infrared pedestrian classification performance of deep neural networks can be effectively improved by the automatic image matting of the infrared pedestrian preprocessing algorithm. The accuracy, precision, and F-value indicators of AlexNet based on automatic image matting of infrared pedestrian preprocessing (i.e., the infrared pedestrian classification algorithm based on automatic image matting) on LSI dataset, the accuracy, precision, and F-value indicators on RIFIR dataset, and the accuracy indicators on KAIST dataset are better than all other algorithms. The VGG based on automatic image matting of infrared pedestrian preprocessing has better accuracy indicators on RIFIR dataset and better recall, precision, and F-value indicators on KAIST dataset than other participating experimental algorithms. Although the performance of the infrared pedestrian classification algorithm based on automatic image matting enhancement on some performance indicators in KAIST dataset is not the best, the performance difference between it and the optimal VGG based on automatic image matting of infrared pedestrian preprocessing is small in these indicators.

1.4.2.3.4 Comparison experiment on preprocessing effect of pedestrian detection in infrared images based on automatic image matting

To verify the preprocessing effect of the introduced automatic image matting algorithm for infrared pedestrian images, a comparison was made between the introduced algorithm and existing infrared pedestrian enhancement methods. The experiment used the min-max enhancement [60] and fuzzy enhancement with top-hat transform [62], which are recently introduced infrared pedestrian enhancement methods, as performance benchmarks, and the data used in the experiment came from LSI and RIFIR datasets.

Fig. 1.31 shows the preprocessing results of the automatic infrared pedestrian image matting algorithm and existing infrared pedestrian enhancement algorithms. Fig. 1.31A and E are the regions of interest in the far-infrared images. Fig. 1.31B and F are the preprocessing results of the introduced automatic infrared pedestrian image matting algorithm. Fig. 1.31C and G are the results of the minimal-maximal enhancement [60].

Fig. 1.31D and H are the results of the top-hat transformation-based blur enhancement [62]. The first two rows show examples when the pedestrian area is brighter than the background area, while the third to sixth rows show examples when the pedestrian area is darker. It can be observed from Fig. 1.31 that when the pedestrian area is relatively bright in the far-infrared images, all three preprocessing algorithms can provide clear pedestrian outlines. When the pedestrian area is relatively dark, the introduced preprocessing algorithm can still provide clear pedestrian outlines and remove cluttered backgrounds, while the other two algorithms are unable to effectively enhance the pedestrian features and produce results with cluttered backgrounds, making it difficult to distinguish pedestrians from the background. These experimental results explain why the introduced automatic infrared pedestrian image matting algorithm can effectively improve the performance of deep neural networks in infrared pedestrian classification tasks, while other preprocessing algorithms cannot.

The purpose of the following experiment was to measure the cost of performance improvement in infrared pedestrian classification for the automatic matting-based deep learning approach. This experiment was conducted on

(A)   (B)   (C)   (D)   (A)   (B)   (C)   (D)

Figure 1.31 Comparison of the preprocessing results of the automatic image matting-based infrared image pedestrian preprocessing algorithm with two existing infrared image pedestrian enhancement methods. (A) Infrared images. (B) Preprocessing results obtained by the proposed approach. (C) Preprocessing results obtained by min-max enhancement [60]. (D) Preprocessing results obtained by fuzzy enhancement with top-hat transform [62]. The first two rows show the cases where pedestrians appear bright. The third to sixth rows show the cases where pedestrians appear dark.

1000 images randomly selected from the test set of LSI dataset. Table 1.18 presents the computational costs of three infrared pedestrian classification algorithms based on infrared image preprocessing using automatic pedestrian matting and the corresponding conventional algorithm in terms of the maximum memory utilization and time consumption.

The proposed approach and conventional AlexNet consumed the least memory in this comparison. The maximum memory consumption of the proposed approach did not increase because the introduced preprocessing and the deep neural network prediction were executed serially and the preprocessing occupied only 2.1 MB of memory. Moreover, the time consumption of the proposed approach and conventional AlexNet is significantly less than other approaches. Compared with AlexNet, the average processing time of the infrared image pedestrian preprocessing based on automatic matting with AlexNet increased by less than 0.02 s. Experimental results show that the infrared image pedestrian preprocessing algorithm based on automatic matting has made significant progress in pedestrian classification performance with very low computational cost. It is worth noting that the infrared image pedestrian classification algorithm

Table 1.18 Comparison of the computation time and maximum memory consumption of the classical deep neural network classification algorithm and the deep neural network classification algorithm based on automatic image matting for pedestrian preprocessing of infrared images.

Infrared pedestrian classification algorithm	Maximum memory consumption (MB)	Time consumption (s)
AlexNet based on infrared image preprocessing using automatic pedestrian matting	1051	0.262
VGG based on infrared image preprocessing using automatic pedestrian matting	1676	1.454
ResNet based on infrared image preprocessing using automatic pedestrian matting	1542	1.175
AlexNet [45]	1051	0.243
VGG [53]	1676	1.435
ResNet [54]	1542	1.153
AlexNet based on infrared image preprocessing using automatic pedestrian matting (on GPU)	1044	0.025

enhanced by automatic matting can achieve a classification speed of 40 frames per second after using GPU acceleration (NVIDIA GTX 1080 graphics card was used in the GPU acceleration experiment), which fully meets the needs of real-time applications.

1.4.2.3.5 Limitations of infrared image pedestrian classification method based on automatic image matting

However, there are also some shortcomings in the infrared pedestrian classification algorithm based on automatic image matting enhancement. Fig. 1.32 shows some examples of classification errors in the infrared pedestrian classification algorithm based on automatic image matting enhancement. Fig. 1.32A and D are positive samples, Fig. 1.32G is a negative sample, Fig. 1.32B, E, and H are the trimaps generated by the infrared image pedestrian trimap automatic generation algorithm in Fig. 1.32A, D, and G, and Fig. 1.32C, F, and I are the alpha mattes obtained by the infrared image pedestrian classification algorithm based on automatic image matting enhancement in Fig. 1.32A, D, and G. The main reason why the infrared pedestrian classification algorithm based on automatic image matting enhancement fails in these examples is that these complex examples do not meet the basic assumptions of the infrared pedestrian trimap automatic generation algorithm: a positive pedestrian sample only has one upright walking pedestrian with a brighter head region than the background region. There are two pedestrians in Fig. 1.32A, and one pedestrian's head region is incorrectly marked as a known background region in the generated trimap of the infrared pedestrian trimap automatic generation algorithm, as shown in Fig. 1.32B. In Fig. 1.32E, due to the deviation of head position, the head region of the

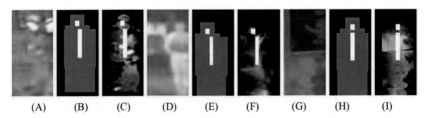

(A)   (B)   (C)   (D)   (E)   (F)   (G)   (H)   (I)

Figure 1.32 Example of misclassification of infrared image pedestrian classification algorithm based on automatic image matting enhancement. (A) and (D) Positive infrared image. (B) and (E) Automatically generated trimap with respect to (A) and (D), respectively. (C) and (F) Alpha mattes regarding (A) and (D), respectively. (G) Negative infrared image. (H) Automatically generated trimap with respect to (G). (I) Alpha mattes regarding (G).

pedestrian in the generated trimap of the infrared pedestrian trimap automatic generation algorithm is incorrectly marked as a known background region, while the object in the background is marked as a foreground region, as shown in Fig. 1.32F. In the trimap corresponding to the third example (as shown in Fig. 1.32H), the purple, pink, and red objects belonging to the background are partially marked as known foreground and partially marked as known background. The incorrect labeling of the trimap results in that the obtained alpha mattes cannot accurately describe the contour of the foreground object (as shown in Fig. 1.32C, F, and I), resulting in classifier prediction errors.

1.4.3 Summary

This section discusses the problem of automatic generation of trimap using the infrared image pedestrian classification task as an example, explores the research ideas of automatic generation of trimap for pedestrians based on expert knowledge in the field of infrared image pedestrian classification, and expands the application prospects of natural image matting technology in the field of infrared image pedestrian classification. An automatic trimap generation method based on human head and upper body positioning was introduced to achieve automatic preprocessing of infrared pedestrians based on the natural image matting technology. This method was combined with a data-driven infrared image pedestrian classification algorithm, obtaining the infrared image pedestrian classification algorithm based on automatic pedestrian matting. Experimental results show that the infrared image pedestrian preprocessing based on automatic pedestrian matting can effectively enhance the contour features of the foreground and suppress irrelevant backgrounds, providing consistent and prominent contour features for the pedestrian target. The infrared image pedestrian classification algorithm based on automatic pedestrian matting outperforms existing data-driven infrared image pedestrian classification algorithms and expert-driven pedestrian classification algorithms on almost all performance indicators in multiple datasets.

In addition, by using the alpha matte obtained from the infrared image pedestrian preprocessing based on automatic image matting as the input data for training, existing advanced deep neural networks can significantly improve their performance on infrared image pedestrian classification tasks, and the computational overhead of the preprocessing is very low. A low-cost solution is provided to improve the performance of deep neural networks by providing consistent pedestrian contour features through image matting based on automatic image matting enhancement.

References

[1] Huang H, Liang Y, Yang X, et al. Pixel-level discrete multiobjective sampling for image matting. IEEE Transactions on Image Processing 2019;28(8):3739−51.

[2] Wang J, Cohen MF. Optimized color sampling for robust matting. In: IEEE Conference on Computer Vision and Pattern Recognition, 2007. 1−8.

[3] He K., Rhemann C., Rother C., et al. A global sampling method for alpha matting. In: IEEE Conference on Computer Vision and Pattern Recognition, 2011. 2049−56.

[4] Feng X, Liang X, Zhang Z. A cluster sampling method for image matting via sparse coding. In: European Conference on Computer Vision, 2016. 204−219.

[5] Shahrian E, Rajan D. Weighted color and texture sample selection for image matting. In: IEEE Conference on Computer Vision and Pattern Recognition, 2012. 718−25.

[6] Shahrian E, Rajan D, Price B, et al. Improving image matting using comprehensive sampling sets. In: IEEE Conference on Computer Vision and Pattern Recognition, 2013. 636−43.

[7] Karacan L, Erdem A, Erdem E. Alpha matting with kl-divergence-based sparse sampling. IEEE Transactions on Image Processing 2017;26(9):4523−36.

[8] Johnson J, Varnousfaderani ES, Cholakkal H, et al. Sparse coding for alpha matting. IEEE Transactions on Image Processing 2016;25(7):3032−43.

[9] Chomicki J, Godfrey P, Gryz J, et al. Skyline with presorting. In: IEEE International Conference on Data Engineering, 2003. 717−19.

[10] Rhemann C, Rother C, Wang J, et al. A perceptually motivated online benchmark for image matting. In: IEEE Conference on Computer Vision and Pattern Recognition, 2009. 1826−33.

[11] Achanta R, Shaji A, Smith K, et al. SLIC superpixels compared to state-of-the-art superpixel methods. IEEE Transactions on Pattern Analysis and Machine Intelligence 2012;34(11):2274−82.

[12] Cao G, Li J, He Z, et al. Divide and Conquer: A Self-Adaptive Approach for High-Resolution Image Matting. In: International Conference on Virtual Reality and Visualization, 2016. 24−30.

[13] Chen Q, Li D, Tang CK. KNN matting. IEEE Transactions on Pattern Analysis and Machine Intelligence 2013;35(9):2175−88.

[14] Gastal ES, Oliveira M. Shared sampling for real-time alpha matting. In: Computer Graphics Forum 2010;29:575−84.

[15] Yao GL. A survey on pre-processing in image matting. Journal of Computer Science and Technology 2017;32(1):122−38.

[16] Cai ZQ, Lv L, Huang H, et al. Improving sampling-based image matting with cooperative coevolution differential evolution algorithm. Soft Computing 2017;21(15): 4417−30.

[17] Knowles JD, Watson RA, Corne DW. Reducing local optima in single-objective problems by multiobjectivization. In: International Conference on Evolutionary Multi-Criterion Optimization, 2001. 269−83.

[18] Greiner D, Emperador JM, Winter G, et al. Improving computational mechanics optimum design using helper objectives: an application in frame bar structures In: International Conference on Evolutionary Multi-Criterion Optimization 2007. 575−89.

[19] Acilar AM, Arslan A. Optimization of multiple input−output fuzzy membership functions using clonal selection algorithm. Expert Systems with Applications 2011;38(3):1374−81.

[20] Wang W, Liu X. Intuitionistic fuzzy information aggregation using Einstein operations. IEEE Transactions on Fuzzy Systems 2012;20(5):923−38.

[21] Liang Y, Huang H, Cai Z, et al. Multiobjective evolutionary optimization based on fuzzy multicriteria evaluation and decomposition for image matting. IEEE Transactions on Fuzzy Systems 2019;27(5):1100−11.

[22] Levin A, Lischinski D, Weiss Y. A closed-form solution to natural image matting. IEEE Transactions on Pattern Analysis and Machine Intelligence 2008;30(2):228−42.

[23] Zhu Q, Shao L, Li X, et al. Targeting accurate object extraction from an image: A comprehensive study of natural image matting. IEEE Transactions on Neural Networks and Learning Systems 2015;26(2):185−207.

[24] Deb K, Pratap A, Agarwal S, et al. A fast and elitist multiobjective genetic algorithm: NSGA-II. IEEE Transactions on Evolutionary Computation 2002;6(2):182−97.

[25] Zhang Q, Li H. MOEA/D: a multiobjective evolutionary algorithm based on decomposition. IEEE Transactions on Evolutionary Computation 2007;11(6):712−31.

[26] Coello C, Lechuga MS. MOPSO: a proposal for multiple objective particle swarm optimization. In: IEEE Congress on Evolutionary Computation 2002;2:1051−6.

[27] Omidvar MN, Yang M, Mei Y, et al. DG2: a faster and more accurate differential grouping for large-scale black-box optimization. IEEE Transactions on Evolutionary Computation 2017;21(6):929−42.

[28] Cheng R, Jin Y. A competitive swarm optimizer for large scale optimization. IEEE Transactions on Cybernetics 2015;45(2):191−204.

[29] Li X, Yao X. Cooperatively coevolving particle swarms for large scale optimization. IEEE Transactions on Evolutionary Computation 2012;16(2):210−24.

[30] Aksoy Y., Ozan Aydin T., Pollefeys M. Designing effective inter-pixel information flow for natural image matting. In: IEEE Conference on Computer Vision and Pattern Recognition, 2017. 29−37.

[31] Li C, Wang P, Zhu X, et al. Three-layer graph framework with the sumD feature for alpha matting. Computer Vision and Image Understanding 2017;162:34−45.

[32] Lee Y, Yang S. Parallel block sequential closed-form matting with fan-shaped partitions. IEEE Transactions on Image Processing 2017;27(2):594−605.

[33] Roychowdhury S, Koozekanani D, Parhi K. Blood vessel segmentation of fundus images by major vessel extraction and subimage classification. IEEE journal of biomedical and health informatics 2014;19(3):1118−28.

[34] Hoover A, Kouznetsova V, Goldbaum M. Locating blood vessels in retinal images by piecewise threshold probing of a matched filter response. IEEE Transactions on Medical Imaging 2000;19(3):203−10.

[35] Zhao X, He Z, Zhang S, et al. Robust pedestrian detection in thermal infrared imagery using a shape distribution histogram feature and modified sparse representation classification. Pattern Recognition 2015;48(6):1947−60.

[36] Lee YS, Chan YM, Fu LC, et al. Near-infrared-based nighttime pedestrian detection using grouped part models. IEEE Transactions on Intelligent Transportation Systems 2015;16(4):1929−40.

[37] Besbes B, Rogozan A, Rus AM, et al. Pedestrian detection in far-infrared daytime images using a hierarchical codebook of SURF. Sensors 2015;15(4):8570−94.

[38] Suard F, Rakotomamonjy A, Bensrhair A, et al. Pedestrian detection using infrared images and histograms of oriented gradients. In: Intelligent Vehicles Symposium, 2006. 206−12.

[39] Wang H, Wang J. An effective image representation method using kernel classification. In: IEEE International Conference on Tools with Artificial Intelligence, 2014. 853−58.

[40] Chien JC, Lee JD, Chen CM, et al. An integrated driver warning system for driver and pedestrian safety. Applied Soft Computing 2013;13(11):4413−27.

[41] Dalal N, Triggs B. Histograms of oriented gradients for human detection. In: IEEE Conference on Computer Vision and Pattern Recognition, 2005. 1: 886−93.

[42] Bay H, Tuytelaars T, Van Gool L. Surf: Speeded up robust features. In: European conference on computer vision, 2006. 404−17.

[43] Kwak JY, Ko BC, Nam JY. Pedestrian tracking using online boosted random ferns learning in far-infrared imagery for safe driving at night. IEEE Transactions on Intelligent Transportation Systems 2017;18(1):69−81.

[44] Collobert R, Weston J, Bottou L, et al. Natural language processing (almost) from scratch. Journal of Machine Learning Research 2011;12(Aug):2493−537.

[45] Krizhevsky A, Sutskever I, Hinton GE. Imagenet classification with deep convolutional neural networks. Communications of the ACM 2012;60(6):84−90.

[46] Simonyan K, Zisserman A. Very deep convolutional networks for large-scale image recognition. arXiv preprint arXiv 2014;1409:1556.

[47] He K, Zhang X, Ren S, et al. Deep residual learning for image recognition. In: IEEE Conference on Computer Vision and Pattern Recognition, 2016. 770−78.

[48] Liang Y, Huang H, Cai Z, et al. Deep infrared pedestrian classification based on automatic image matting. Applied Soft Computing 2019;77:484−96.

[49] Lee MW, Cohen I. A model-based approach for estimating human 3D poses in static images. IEEE Transactions on Pattern Analysis and Machine Intelligence 2006;28 (6):905−16.

[50] Rhemann C, Rother C, Rav-Acha A, et al. High resolution matting via interactive trimap segmentation. In: IEEE Conference on Computer Vision and Pattern Recognition, 2008. 1−8.

[51] Jia Y, Shelhamer E, Donahue J, et al. Caffe: Convolutional architecture for fast feature embedding. In: ACM International Conference on Multimedia, 2014. 675−78.

[52] Khellal A, Ma H, Fei Q. Pedestrian Classification and Detection in Far Infrared Images. In: International Conference on Intelligent Robotics and Applications, 2015. 511−22.

[53] Miron A, Rogozan A, Ainouz S, et al. An evaluation of the pedestrian classification in a multi-domain multi-modality setup. Sensors 2015;15(6):13851−73.

[54] Hwang S, Park J, Kim N, et al. Multispectral pedestrian detection: Benchmark dataset and baseline. In: IEEE Conference on Computer Vision and Pattern Recognition, 2015. 1037−45.

[55] Zhang X, Chao W, Li Z, et al. Multi-modal kernel ridge regression for social image classification. Applied Soft Computing 2018;67:117−25.

[56] Dollar P, Wojek C, Schiele B, et al. Pedestrian detection: An evaluation of the state of the art. IEEE Transactions on Pattern Analysis and Machine Intelligence 2012;34 (4):743−61.

[57] Vailaya A, Figueiredo MA, Jain AK, et al. Image classification for content-based indexing. IEEE Transactions on Image Processing 2001;10(1):117−30.

[58] Wang G, Liao TW. Automatic identification of different types of welding defects in radiographic images. Ndt & E International 2002;35(8):519−28.

[59] Han J, Zhang D, Hu X, et al. Background prior-based salient object detection via deep reconstruction residual. IEEE Transactions on Circuits and Systems for Video Technology 2015;25(8):1309−21.

[60] Kim T, Kim S. Pedestrian detection at night time in FIR domain: comprehensive study about temperature and brightness and new benchmark. Pattern Recognition 2018;79:44−54.

[61] He F, Guo Y, Gao C. An improved pulse coupled neural network with spectral residual for infrared pedestrian segmentation. Infrared Physics & Technology 2017;87:22−30.

[62] Soundrapandiyan R, Mouli PC. Adaptive pedestrian detection in infrared images using fuzzy enhancement and top-hat transform. International Journal of Computational Vision and Robotics 2017;7(1−2):49−67.

CHAPTER 2

Application of intelligent algorithms in the field of logistics planning

2.1 Overview of research progress

With the continuous development of the logistics field, more and more freight vehicles are being utilized for urban logistics transportation. Although this trend has brought great convenience to urban logistics, it has also posed significant challenges to traffic management and environmental protection. To address these challenges, a two-echelon logistics distribution system has been introduced, wherein large trucks transport goods from warehouses to transit stations, and then small trucks distribute the goods for the final stage of transportation. This system could limit the entry of large trucks into the city efficiently, thereby reducing their driving mileage within urban areas. However, constructing a two-echelon logistics distribution system requires solving the two-echelon vehicle routing problem (2E-VRP), which is a nondeterministic polynomial hard (NP-hard) problem more complex than the classical routing problem. As a complex planning problem at the forefront of urban logistics, the large-scale 2E-VRP has significant research significance and academic value. Intelligent Algorithm Laboratory at School of Software Engineering, South China University of Technology has introduced a graph-based fuzzy evolutionary algorithm (GFEA) [1] to solve the 2E-VRP, drawing on the advantages of current algorithms. Starting from fuzzy correlations between users, this algorithm addresses the complexity of the large-scale routing optimization problem in two-echelon distribution. It aims to improve the solution performance of two-echelon vehicle routing problems and provide an efficient and feasible implementation plan for multiechelon transportation scheduling in urban logistics. Below are some application cases based on the strategies proposed by this algorithm.

Example 1 presents a demonstration software for logistics multiechelon scheduling, as depicted in Fig. 2.1. This case combines intelligent algorithms to solve practical problems, such as batching plant selection, one-echelon

Intelligent Algorithms
DOI: https://doi.org/10.1016/B978-0-443-21758-6.00003-6

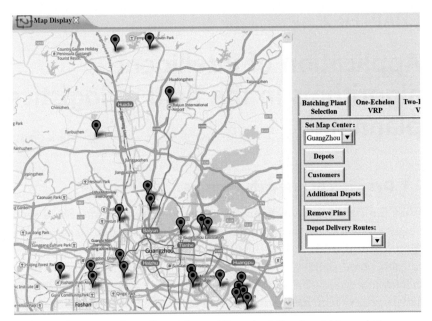

Figure 2.1 Demonstration software for multilayer logistics dispatching solution.

vehicle routing problems, and two–echelon vehicle routing problems. The software has been successfully applied in the planning and design schemes of important cities, including Guangzhou and Nanning, significantly improving the quality of planning and the efficiency of planners. Furthermore, it has also been successfully utilized in the cold chain logistics system of well-known logistics companies in China, yielding remarkable results.

Example 2 presents a large-scale port optimization project, as depicted in Fig. 2.2. This project comprehensively applies various information technologies, control technologies, operational management technologies, port resources, and e-commerce technologies to achieve interconnection and sharing of human resources, funds, materials, equipment, and other resources in the port. Through the idea of using GFEA to solve the 2E-VRP problem, it enables full-scale management of the port with horizontal connectivity and vertical control.

These examples illustrate that GFEA has certain advantages in solving large-scale path optimization problems by avoiding the issue of different transit station changes due to unreasonable allocation relationships and reducing the number of estimations of the evaluation function. Moreover, the fuzzy allocation graph estimated from the information exchange of

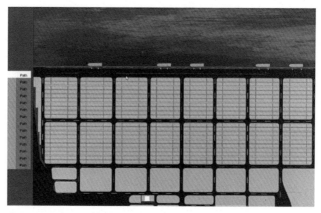

Figure 2.2 Demonstration of large port optimization project.

different individuals in each generation of the population maintains the advantage of the population during the iterative allocation learning process. However, GFEA still requires further improvement. For instance, the graph-based fuzzy operator is sensitive to the number of fuzzy subsets, and the fuzzy allocation indicators that affect the evaluation of the solution also necessitate further discussion.

2.2 Scientific principles

2.2.1 Problem description

In the intelligent distribution process, goods are not directly delivered to the target customers but are transferred through satellite stations as temporary warehouses. First, vehicles with strong load capacity transport the goods to the satellite stations. Then, small vehicles deliver the goods to the target customers from the satellite station, as shown in Fig. 2.3.

In a weighted undirected graph $G(V, E)$, V represents the vertex set of the graph, and E represents the edge set of the graph. $V = V_d \cup V_s \cup V_c$, where V_d is the warehouse vertex set, V_s is the satellite station vertex set, and V_c is the customer vertex set. In the two-echelon vehicle routing optimization problem, the number of vertices in the warehouse set is 1, the number of vertices in the satellite station set is m, and the number of vertices in the customer set is n. $E = E_1 \cup E_2$, where the first-echelon path E_1 and the second-echelon path E_2 divide the edge set of graph G into two parts. $E_1 = \{\{i,j\}:i<j, i,j \in \{v_0\} \cup V_s\}$. The first-echelon path includes the path between the warehouse and the satellite station and the path between satellite

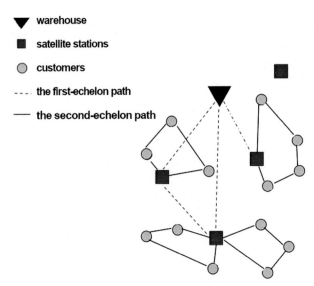

Figure 2.3 Example of feasible solution for two-echelon vehicle routing optimization problem.

stations. The demand of satellite station vertices on the first-echelon path can be split. $E_2 = \{\{i,j\}:i<j, i,j \in V_s \cup V_c, \{i,j\} \notin V_s \times V_s\}$. The second-echelon path includes the path between the satellite station and the customer and the path between customers. On the second-echelon path, the weight of the path between customer vertex i and customer vertex j is c_{ij}. The demand of each customer is d_i, and the service must be completed in a single attempt. The main objective of the two-echelon vehicle routing optimization problem is to serve all customers while satisfying the capacity constraints of the vehicles in each echelon. This could minimize the combined weight of both the first-echelon and second-echelon paths. The mathematical model for this is as follows:

$$\min \sum_{r \in R_1 \cup R_2} p_r x_r \tag{2.1}$$

$$\sum_{r \in R_1, s \in V_s} q_{rs} \leq Q_1 x_r \tag{2.2}$$

$$d_r \leq Q_2 \tag{2.3}$$

$$\sum_{r \in R_1, s \in V_s} q_{rs} = \sum_{r \in R_2, s \in V_s} d_r x_r S_{rs} \tag{2.4}$$

$$\sum_{r \in R_2} \text{visited}_{rs} x_r = 1 \tag{2.5}$$

$$\sum_{r \in R_1} x_r \leq b_1 \tag{2.6}$$

$$\sum_{r \in R_2, c \in V_s} x_r S_{rc} \leq b_2 \tag{2.7}$$

$$q_{rs} \geq 0 \, r \in R_1, s \in V_s \tag{2.8}$$

$$0 \leq s \leq m, 0 \leq c \leq n \tag{2.9}$$

$$x_r \in \{0, 1\} \quad r \in R_1 \cup R_2 \tag{2.10}$$

$$S_{rs} \in \{0, 1\} \quad r \in R_2, s \in V_s \tag{2.11}$$

$$\text{visited}_{rc} \in \{0, 1\} \quad r \in R_2, c \in V_c \tag{2.12}$$

In Eq. (2.1), pr is the weight of r, R_1 is all the paths on the first-echelon path, and R_2 is all the paths on the second-echelon path. Therefore, for any path $r \in R_1 \cup R_2$, we define variable x_r:

$$x_r = \begin{cases} 1 & \text{If the path } r \text{ is included in the feasible solution} \\ 0 & \text{Otherwise} \end{cases} \tag{2.13}$$

Equations (2.2) and (2.3) indicate that the capacity of each path r on the first-echelon and second-echelon paths cannot exceed the vehicle's load capacity, respectively. Q_1 and Q_2 are the capacities of the vehicles on the first- and second-echelon paths respectively, q_{rs} is the load required by satellite station node s on the first-echelon path, and $d_r = \sum_{c \in V_c : r \in R_2} d_c$ is the load of satellite station node on the first-echelon path. Eq. (2.4) indicates that for a given satellite station node, the amount of goods on the first echelon should be consistent with the amount of goods on the second echelon. The variable S_{rs} is defined as follows:

$$S_{rs} = \begin{cases} 1 & \text{If both the strating and ending nodes of path } r \text{ are satellite nodes } s \\ 0 & \text{Otherwise} \end{cases}$$

$$\tag{2.14}$$

Eq. (2.5) ensures that each customer is served only once. Equation (2.6) and (2.7) indicate that the number of vehicles on the first-echelon and second-echelon paths cannot exceed the limits, and b_1 and b_2 are the available number of vehicles on the first-echelon and second-echelon paths, respectively. Equations (2.8)−(2.12) denote the restriction on the range of variable values.

It is evident that the first-echelon path in the two-echelon vehicle routing optimization problem is determined by the second-echelon path based on the above definition. This can be simply understood as two allocation subproblems: warehouse-satellite and satellite-customer. In essence, the warehouse, satellite station, and customers are treated as three sets of heterogeneous objects, where the satellite station is seen as the central object. The two-echelon vehicle routing optimization problem is an optimization problem of a special compatible bipartite graph structure.

2.2.2 Problem analysis

As shown in Fig. 2.3, 2E-VRP can be represented by a weighted undirected graph $G(V, E)$, where V and E represent the vertex set and edge set, respectively. The vertex set V consists of the warehouse node V_0, a subset V_S of m transit stations, and a subset V_c of n customers. In the edge set $E = E_1 \cup E_2$, where E_1 is the edge subset composed by the first-echelon paths and E_2 is the edge subset composed by the second-echelon paths. The 2E-VRP problem can be understood as the overall plan of designing the first-echelon and second-echelon routes under given constraint conditions. Due to variations in the number of vehicles and their load capacities in different transportation echelons, the number of VRP subproblems on the second-echelon path remains uncertain. It is challenging to completely segregate the VRP subproblems of the first and second echelons using only distance information or other simple data. This presents a significant hurdle to research on developing a two-echelon logistics distribution system. Additionally, as the scale of the 2E-VRP problem increases, the distribution relationship between transit stations and customers becomes increasingly complex.

2.2.3 Algorithm design

To tackle the uncertainty in satellite-customer allocation in the two-echelon vehicle routing optimization problem, this chapter presents a fuzzy evolutionary heuristic allocation method that focuses on the satellite-customer matching problem on the second-echelon path. The demand of a satellite station on the first-echelon path is the sum of demands of the customers served by that satellite station on the second-echelon path. Changes in the satellite-customer allocation on the second-echelon path would directly impact the routing arrangement on the first-echelon path. Furthermore, the allocation of satellite stations to customers breaks down the two-echelon vehicle routing optimization problem into several path optimization subproblems. The partitioning of the path optimization subproblems is influenced by the correlation between customers who are served by the same satellite station or on the same path. This correlation among customers has a direct impact on the satellite-customer allocation problem, which is the solution to the path optimization subproblems in the two-echelon vehicle routing optimization problem. To describe the effect of customer correlation on satellite station allocation, this section employs fuzzy set correlation theory to decompose the relationship among customer sets. In other words, the fuzzy allocation process is used to solve the subproblems in the two-echelon vehicle routing optimization problem, as depicted in Fig. 2.4.

The fuzzy evolutionary algorithm based on graphs is shown in Fig. 2.5. In the initialization step, N initial solutions are randomly generated. Following this, the fuzzy subsets are obtained using the fuzzy decomposition algorithm. The fuzzy allocation strategy first utilizes the matching between the fuzzy subset and the satellite station as heuristic information to obtain the second-echelon path of the two-echelon vehicle routing optimization problem. The fuzzy satellite-allocation process is then used to update the offspring population.

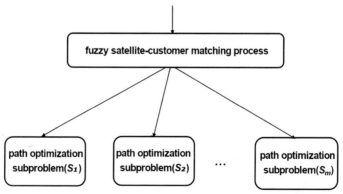

Figure 2.4 Main idea of using the fuzzy evolutionary heuristic algorithm to solve the two-echelon vehicle routing optimization problem.

Algorithm 2-1 Basic framework of fuzzy evolutionary algorithm based on graph

---

Input: Test data for two-echelon path problem: number of fuzzy subsets N_{cf},
population size N, customer set V_c, transit station set V_s

Output: Optimal solution s_r

   // Initialization
   1) $s_r = \emptyset$.
   2) Randomly construct the initial population pop and generate the allocation graph G^*.
   // Fuzzy allocation strategy:
   while   termination conditions are not satisfied   **do**
   1) Select $2/N$ better individuals from the population pop to obtain the fuzzy equivalence matrix R for customer and transit station distribution.
   2) Decompose the fuzzy equivalence matrix R to obtain fuzzy subsets A_k, $k = 1, 2, \ldots, N_{cf}$.
   3) Match the transit station set V_s with the fuzzy subset A_k.
   4) Set $i = 1$.
   5) For each population individual p_i ($p_i \in pop$), use the fuzzy local search strategy based on fizzy subsets and allocation graph G^* to obtain the fuzzy matching weights of each transit station node and customer node.
   6) Use the fuzzy local search strategy to update the second echelon, and use the fuzzy neighborhood selection strategy to construct the first-echelon path of the population individual to obtain the new individual p_i'.
   7) Update S_r with p_i' if $C(p_i') > C(s_r)$.
   8) Set $i \leftarrow i + 1$. If $i \leq N$. Retum to step (4).
   9) Sort the individuals in the population pop and select $N/2$ better individuals, and randomly generate $N/2$ new individuals to mergeand update the population pop and allocation graph G^*.
   end while

Figure 2.5 Process of graph-based fuzzy evolutionary algorithm.

The graph–based fuzzy evolutionary algorithm begins by constructing the initial population and utilizing the statistical fuzzy estimation method to create a fuzzy allocation graph of the current population. The allocation relationship of the two–echelon routes in the fuzzy allocation graph is updated iteratively. Different fuzzy heuristic operators are designed to generate new solutions for the next generation. Ultimately, a novel method is developed to solve the two–echelon path allocation problem with large-scale customer demands. Fig. 2.6 depicts the population fuzzy allocation graph constructed using the statistical fuzzy estimation method.

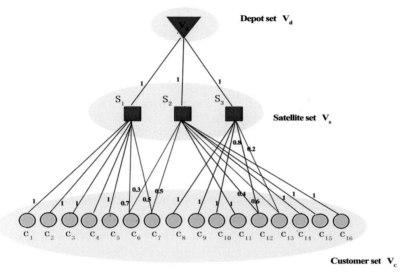

Figure 2.6 Fuzzy allocation graph of the population.

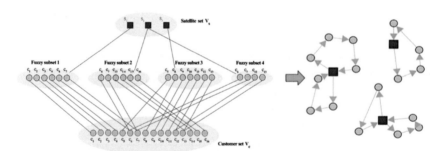

Figure 2.7 Fuzzy user subsets with specific relationships.

To enhance the search efficiency of the first-echelon path construction, a fuzzy neighborhood search strategy based on the probability graph is employed to find a more effective search path.

The fuzzy relationship matrix among the current population of users is decomposed to obtain a series of fuzzy user subsets with specific relationships based on the fuzzy allocation graph, as shown in Fig. 2.7. This algorithm utilizes fuzzy subsets as heuristic information and designs corresponding fuzzy operator algorithms to obtain the second-echelon optimized path for the 2E-VRP problem.

2.2.4 Algorithm analysis

To analyze the feasibility of the fuzzy allocation heuristic strategy for solving the two-echelon vehicle routing optimization problem, the fuzzy allocation performance coefficient Fa is defined as shown below to evaluate the contribution of fuzzy subsets and satellite station matching to the problem. For each individual p_i in the population, Fa is defined as

$$Fa = \sum_{i=1}^{n_s} F_{p_i} \qquad (2.15)$$

where p_i is each individual in the population, F_{p_i} is the average allocation coefficient between each satellite station and its associated fuzzy subset in the second-echelon path of individual p_i, and n_s refers to the number of satellite stations. The fuzzy allocation performance indicator is used to measure the impact of fuzzy relationships on satellite-customer allocation in the customer set.

We use the sample correlation coefficient to illustrate the relationship between fuzzy satellite-customer allocation and feasible solution performance. For each test case, we run 30 independent experiments. The vector $\{f_1', ..., f_l'\}$ corresponds to the reciprocal of the optimal solution obtained in each experiment, while the vector $\{Fa_1, ..., Fa_l\}$ corresponds to the fuzzy allocation performance coefficient obtained for each feasible solution. The correlation coefficient ρ between the fuzzy allocation performance vector and the feasible solution correlation vector is as follows:

$$\rho = \frac{\sum_{i=1}^{l}\left(f_i' - \overline{f_i'}\right)\left(Fa_i - \overline{Fa_i}\right)}{\sqrt{\sum_{i=1}^{l}\left(f_i' - \overline{f_i'}\right)^2}\sqrt{\sum_{i=1}^{l}\left(Fa_i - \overline{Fa_i}\right)^2}} \qquad (2.16)$$

where l refers to the number of times the independent experiment is performed, and f_i' is the reciprocal of the feasible solution, which is defined as

$$f_i' = \frac{1}{f_i} \qquad (2.17)$$

Fig. 2.8 shows the box plot of the correlation coefficient ρ for all test cases in four test sets, which indicates the correlation between the fuzzy allocation vector and the optimal solution vector. In Set 2 and Set 3, the value of ρ is generally greater than 0.4, and the change in the weight of the optimal solution is related to the fuzzy allocation value. In Set 4 and

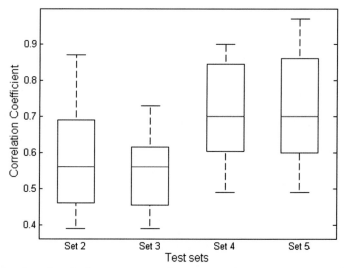

Figure 2.8 Box plot of ρ for test cases on four test sets (Set 2, Set 3, Set 4, and Set 5).

Set 5, the value of ρ is generally greater than 0.6, and there is a close relationship between the change in the weight of the optimal solution and the fuzzy allocation value. In general, the fuzzy allocation value is positively correlated with the reciprocal of the optimal solution, meaning that the larger the fuzzy allocation value, the lower the weight of the optimal solution. For large-scale test sets, where there are more satellite stations and the customer distribution is more complex, the fuzzy satellite-customer allocation process can provide a better solution for the two-echelon vehicle routing optimization problem.

2.2.5 Experimental analysis

To illustrate the performance of the fuzzy evolutionary allocation heuristic algorithm introduced in this chapter on large-scale test cases, Fig. 2.9 shows the box plots of GFEA, GRASP [2], LNS-2E [3], and ALNS [4] on Set 5 instance. The results in Fig. 2.9 show that the feasible solutions obtained by the GFEA algorithm are more stable than the other three algorithms on most test instances in Set 5. The GFEA algorithm obtains better feasible solutions than the GRASP algorithm, mainly because the fuzzy satellite-customer allocation method is more effective than the distance clustering-based allocation method used in the GRASP algorithm. In addition, compared with LNS-2E and ALNS, GFEA can obtain more robust feasible solutions.

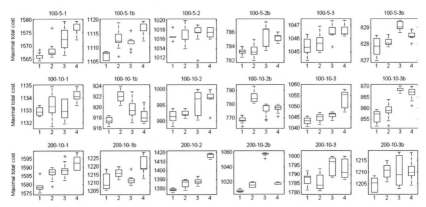

Figure 2.9 Box plots of the worst feasible solutions obtained in 30 independent experiments for the test cases in the Set 5 test set. The 1, 2, 3, and 4 on the x-axis represent the GFEA algorithm, LNS-2E algorithm, GRASP algorithm, and ALNS algorithm, respectively. *GFEA*, Graph-based fuzzy evolutionary algorithm.

Table 2.1 shows the experimental results of different algorithms on the Set 5. The column names "Average," "Std," and "T(s)" respectively display the average weight, standard deviation, and average execution time of 30 independent experiments. In Table 2.1, if the average values of the LNS-2E algorithm, the GRASP algorithm, and the ALNS algorithm in the test cases are bolded, it means that the algorithm is superior to the GFEA algorithm from a significant statistical perspective. If the average values of the LNS-2E algorithm, the GRASP algorithm, and the ALNS algorithm in the test cases are underlined, it indicates that the results of this algorithm are superior to those of the GFEA algorithm from a significant statistical perspective. However, if the average values of the LNS-2E algorithm, the GRASP algorithm, and the ALNS algorithm are not marked with an asterisk, it means that the GFEA algorithm shows no significant difference compared to the other three algorithms.

We compared the GFEA algorithm with the LNS-2E algorithm, GRASP algorithm, and ALNS algorithm using nonparametric estimation by conducting 30 independent runs. Assuming a significance echelon of $\alpha = 0.05$. For instance, in the test case 100−5-1 of Set 5, the average weight of GFEA is 1565.46, the average weight of LNS-2E is 1566.87, the average weight of GRASP is 1569.42, and the average weight of ALNS is 1578.4. Clearly, the statistical performance of GFEA is superior. Thus, the average value is marked with an asterisk, while the average values of the other three algorithms are underlined. In the test case

Table 2.1 Demonstration of the solution weight results on Set 5.

Instances	GFEA			LNS-2E			GRASP			ALNS		
	Average	Std	T(s)	Average	Std	T(s)	Average	Std	T(s)	Average	Std	T(s)
100–5-1	1565.46*	1.80	34	1566.87	0.00	65	1569.42	5.27	140	1578.4	4.28	235
100–5-1b	1108.62*	2.30	32	1111.93	2.13	60	1111.98	3.56	132	1118.95	1.34	155
100–5-2	1016.32*	2.70	41	1017.94	5.27	79	1017.34	6.41	178	1016.36	7.51	183
100–5-2b	783.18	4.90	33	782.25*	3.56	45	786.04	7.92	130	785.02	8.59	130
100–5-3	1045.29*	3.90	20	1045.61	6.41	31	1046.67	9.18	99	1046.17	10.10	124
100–5-3b	828.99	0.01	21	828.54*	7.92	42	829.87	4.92	92	828.99	3.12	99
100–10-1	1132.23	3.98	45	1132.11*	1.18	39	1132.23	12.31	120	1133.17	8.37	169
100–10-1b	917.01*	2.56	64	922.85	4.92	68	917.05	6.53	210	917.35	9.12	205
100–10-2	990.58*	4.09	79	991.61	8.31	92	991.78	1.05	180	997.42	3.02	204
100–10-2b	768.65*	1.28	56	786.66	6.53	103	777.98	4.08	160	773.56	7.14	174
100–10-3	1043.65	3.57	78	1043.55	1.05	78	1043.57*	2.97	98	1055.88	6.03	648
100–10-3b	853.12*	3.05	45	858.72	4.08	89	867.32	4.35	120	863.42	4.39	205
200–10-1	1575.79*	3.01	68	1598.46	6.97	132	1587.12	4.38	187	1697.83	5.01	220
200–10-1b	1209.62*	2.58	78	1217.23	1.35	148	1210.18	5.12	189	1225.61	6.10	189
200–10-2	1375.74*	0.19	89	1376.16	4.38	155	1389.94	4.17	170	1419.94	3.22	173
200–10-2b	1004.15*	1.09	100	1016.05	5.12	145	1057.90	9.01	147	1018.83	5.12	147
200–10-3	1788.03*	0.08	98	1789.44	4.17	210	1792.49	8.24	111	1799.76	4.57	625
200–10-3b	1201.92*	1.26	120	1206.85	8.01	180	1203.61	12.16	164	1208.61	17.36	194
#. of "w-d-l"				12–3–3			11–6–1			13–5–0		
Average	1122.69	2.35	61	1127.38	4.52	97.83	1129.58	6.20	145.94	1138.07	6.46	226.61

GFEA, Graph-based fuzzy evolutionary algorithm.

100−10-3, the statistical performance of the GRASP algorithm (with an average weight of 1043.57) is better than that of other algorithms, so its average value is marked with an asterisk. In this test case, the performance of the GRASP algorithm is better than that of the GFEA algorithm, so it is bolded. The statistical performance of the LNS-2E algorithm and the ALNS algorithm is not superior to that of the GFEA algorithm, so they are not bolded. At the bottom of the table, there are two additional rows.

The row "w-d-l" represents the comparison between GFEA and other algorithms, where w, d, and l indicate the number of test cases in which the performance is better, similar, and worse than a certain algorithm, respectively. The number of test cases with no performance difference and the number of test cases with worse performance than other test cases are not shown. The last row represents the average value of all test cases for a certain performance feature of the algorithm.

It is evident that the GFEA algorithm outperforms the LNS-2E algorithm, the GRASP algorithm, and the ALNS algorithm in most of the test cases based on the results presented in Table 2.1. The GFEA algorithm has several advantages over the ALNS algorithm. Out of the 18 test cases in Set 5, the GFEA algorithm performs better than the ALNS algorithm in 16 cases. For the large-scale Set 5 test set, the GFEA algorithm is slightly inferior to the LNS-2E algorithm in only five test cases (100−5-2b, 100−5-3b, 100−10-1, 100−10-2b, and 100−10-3), but the difference in average weight is not significant. The GFEA algorithm also outperforms the GRASP algorithm based on satellite-customer allocation. Out of the 18 test cases in Set 5, the GFEA algorithm performs better than the GRASP algorithm in 16 cases. Furthermore, the GFEA algorithm is competitive with the current LNS-2E algorithm for solving the two-echelon vehicle routing problem. This is mainly because the fuzzy allocation strategy of the GFEA algorithm reduces the repair and destruction operations between different satellite stations, which results in more stable solutions for the two-echelon vehicle routing problem. Moreover, in the large-scale Set 5 test set, the GFEA algorithm executes faster than the LNS-2E algorithm, mainly because the fuzzy allocation reduces the frequent search process of different satellite stations, thus avoiding unnecessary evaluations.

2.3 Summary

This chapter introduces a fuzzy evolutionary algorithm that utilizes graph structure information to solve the large-scale two-echelon vehicle routing optimization problem. To address the allocation problem between two-echelon paths, a fuzzy operator based on graph structure is incorporated into the iterative learning of the evolutionary algorithm. The heuristic search strategy with the fuzzy allocation graph enhances the evolutionary advantage of the population and reduces redundant search between different transit stations. The population is then updated from the fuzzy allocation graph of parent individuals through the fuzzy allocation process based on graph structure. Finally, the effectiveness of the graph-based fuzzy evolutionary algorithm is demonstrated through experiments on a public test set. For more information and related code, please refer to reference [1] and visit: http://www2.scut.edu.cn/huanghan/fblw/list.htm.

References

[1] Yan X, Huang H, Hao Z, Wang J. A graph-based fuzzy evolutionary algorithm for solving two-echelon vehicle routing problems. IEEE Transactions on Evolutionary Computation 2019;24(1):129−41.
[2] Crainic TG, Mancini S, Perboli G, et al. GRASP with path relinking for the two-echelon vehicle routing problem. Advances in metaheuristics. New York, NY: Springer New York; 2013. p. 113−25.
[3] Breunig U, Schmid V, Hartl RF, et al. A large neighbourhood based heuristic for two-echelon routing problems. Computers & Operations Research 2016;76:208−25.
[4] Hemmelmayr VC, Cordeau JF, Crainic TG. An adaptive large neighborhood search heuristic for two-echelon vehicle routing problems arising in city logistics. Computers & Operations Research 2012;39(12):3215−28.

CHAPTER 3

Application of intelligent algorithms in the field of software testing

3.1 How to solve excessive overhead of software testing—starting with the automated test case generation problem

3.1.1 Overview of research progress

In recent years, in the software development process (as shown in Fig. 3.1), it is estimated that 50% of the overhead comes from the testing process. The main purpose of software testing is to find out various potential defects and errors in software with minimum human, material, and time. Fixing these defects and errors can avoid problems caused by potential issues after software release. Among them, black-box testing and white-box testing are two common types of testing. Black-box testing focuses on evaluating the performance of the test program, while white-box testing can reveal potential defects in the logic of the program. The automated test case generation (ATCG) is a type of white-box testing problem that needs to be solved urgently. In the past, the ATCG was mostly realized by manual means, and the solution of the ATCG problem can effectively reduce the human and

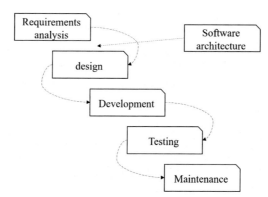

Figure 3.1 Diagram of software development process.

Intelligent Algorithms
DOI: https://doi.org/10.1016/B978-0-443-21758-6.00001-2

material resource expenses in the software testing process. To solve the problem of human, material, and time costs in software testing, the Intelligent Algorithms Laboratory of the School of Software in South China University of Technology has conducted a series of research on the automatic generation of test cases. Currently, the research on the test-case-path relationship matrix for ATCG in IEEE Transactions on Industrial Informatics has been published. The following is a review of the development process and results of this work.

First, this work introduces the background of the automated test case generation for path coverage (ATCG-PC) problem in the fog computing program. On the premise of understanding the concept of the whole fog computing system (as shown in Fig. 3.2), we can know that the goal of the ATCG-PC problem in the fog computing program is to maximize the path coverage. At the same time, the test case overhead needs to be controlled to the minimum possible. In solving the ATCG problem, the fog computing program is only used as a benchmark function for comparison. The requirement for ATCG based on path coverage is to find test cases that cover all feasible paths in the benchmark function within a limited test case overhead. Then, for the problem that there are still some paths that cannot be covered by test cases, we introduce a differential evolution based on relationship programming (RP-DE) algorithm based on the test-case–path relationship matrix.

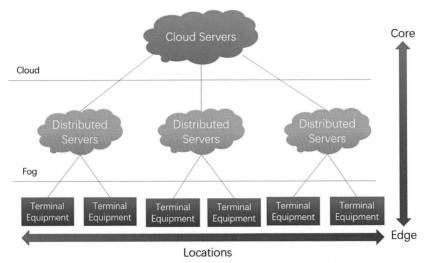

Figure 3.2 Diagram of fog computing system.

The RP-DE algorithm identifies the homogeneous low-dimensional Euclidean space, which is the search space composed of the test case encoding dimensions related to the target path. It is updated by collecting the test case variables and path node correlations in the test case encoding space. The algorithm is guided by the information in the relationship matrix and can allocate more computational resources for searching this homogeneous low-dimensional Euclidean space. It can reduce the test case overhead of the algorithm and improve the search efficiency of the algorithm.

The process of RP-DE is mainly to initialize the population and the relationship matrix R. Then, it repeats the mutation and crossover operations of the DE algorithm and the target path will be covered based on the test-case-path relationship matrix until the final termination condition is met. When the path coverage c is 100% or the number of generated test cases T is greater than or equal to the preset maximum test case overhead (Max), the algorithm stops and exits. The algorithm framework is shown in Fig. 3.3.

This section has successfully implemented the generation of unit test cases for the fog computing toolkit iFogSim based on several functions commonly used as benchmark test functions and the above-mentioned algorithmic ideas.

In this section, we introduce a mathematical model of the ATCG-PC problem for fog computing programs, which is a single-objective optimization problem that needs to be evaluated only once per test case. This model solves the problem that other mathematical models cannot compare the performance of solution algorithms when fog computing programs and other programs have infeasible paths. We also introduce the RP-DE algorithm, which has a significant advantage over the compared algorithms in unit tests for testing fog computing programs and other test programs, and has strong robustness (Figs. 3.4 and 3.5).

3.1.2 Scientific principle
3.1.2.1 Problem description
This section presents the mathematical model of the ATCG-PC problem which aims to maximize the objective coverage within a certain overhead. Given a system under test, $X = \{x_1, x_2, ..., x_N\}$ denotes the set of candidate solutions in the decision space, where N is a large number.

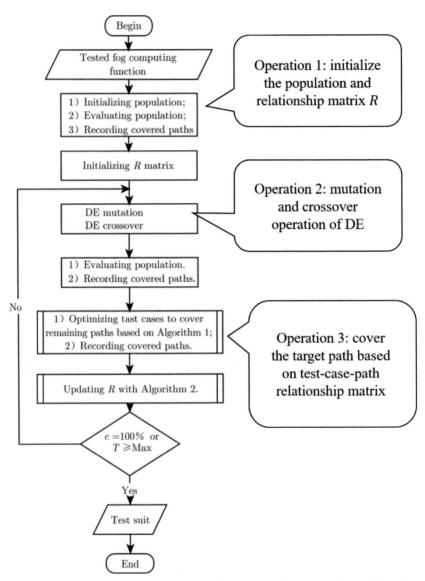

Figure 3.3 Framework of RP-DE. *RP-DE*, Differential evolution based on relationship programming.

The overhead of traversing all candidate solutions is too large. $P_0 = \{p_1, p_2, ..., p_L\}$ denotes the set of paths and T denotes the number of test case evaluations. *Max* denotes the maximum number of evaluations. ATCG–PC problem can be modeled as follows.

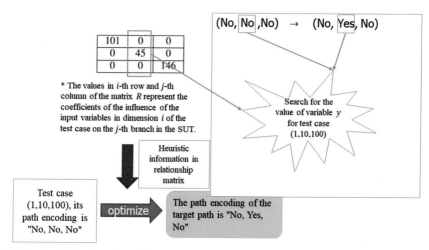

Figure 3.4 Diagram of updating test cases based on relationship matrix (1).

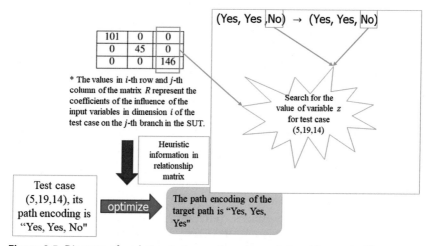

Figure 3.5 Diagram of updating test cases based on relationship matrix (2).

Maximize c (path coverage rate)
subject to

$$c = \frac{l}{L} \times 100\% \tag{3.1}$$

$$l = \sum_{j=1}^{L} \min\left\{ 1, \sum_{i=1}^{T} \theta_{ij} \right\} \tag{3.2}$$

$$T \leq \text{Max} \tag{3.3}$$

$$\theta_{ij} = \begin{cases} 1, & \text{testcase } X_i \text{ covers path } p_j, \\ 0, & \text{otherwise} \end{cases} \tag{3.4}$$

$$\sum_j^L \theta_{ij} = 1 \tag{3.5}$$

$$X_i = \{x_{i,1}, \ldots, x_{i,k}, x_{i,k+1}, \ldots, x_{i,n}\}$$

$$x_{i,j} \in Z; \; lb_j \leq x_{i,j} \leq lb_j$$

$$x_{i,j'} \in R, \; lb_{j'} \leq x_{i,j'} \leq lb_{j'}; \; 1 < i \leq N; \; j = 1, 2, \ldots, k; \; j' = k+1, k+2, \ldots, n \tag{3.6}$$

where Constraints (3.1) and (3.2) define variable c which represents the proportion of covered paths to the total paths. Constraints (3.3) and (3.4) use the intermediate variable θ_{ij} to define the path coverage condition in which each test case covers one and only one path. Constraint (3.5) is a constraint on the search overhead of the algorithm, where the maximum number of fitness evaluations of the problem is less than a predefined maximum Max. Constraint (3.6) defines a constraint on the range of decision space values and variable types for the ATCG-PC problem. Since the test cases may be composed of integers, floating point numbers, or even strings and arrays, the model treats them uniformly as arrays of integers and floating-point numbers within a certain length.

There are three difficulties in solving the ATCG-PC problem. First, the problem is a nondeterministic polynomial hard (NP-hard) problem with a large-scale decision space. Traversing all feasible test cases in the search space is not acceptable. Second, the encoding relationship between the input test cases and the output paths is not clear. Besides, ATCG-PC is not continuously differentiable. Therefore, this problem cannot be solved using the CPLEX tool or other convex optimization algorithms. Third, the ATCG-PC problem requires different paths to be covered depending on the program under test. Some of these paths are more easily covered or can be covered first. The third difficulty in solving this problem is how to use the information of the covered paths to guide the subsequent search of the algorithm and approximately reduce the repeated calculations.

3.1.2.2 Differential evolution based on test-case-path relationship matrix for automated test case generation

The following is an example of the DE algorithm. The framework of the DE algorithm based on the test-case-path relationship matrix (RP-DE) is shown below.

Operation 1: Initialize the population, evaluate the population, and record the path coverage.

Operation 2: After updating the population with DE's crossover, variation, and selection operators, evaluate the population and record the path coverage.

Operation 3: Use the test-case-path relationship matrix to guide the algorithm to cover the remaining paths, and update the relationship matrix.

In Operation 3, the test-case-path relationship matrix is used to optimize test case x_{old} to cover the target path p_{target}. It includes three specific operations, namely, selecting the target path p_{target}, generating test cases based on the information of the relationship matrix R, and updating the relationship matrix R.

In the first step of selecting the target path p_{target}, the path encoding of the path p_{x_i} covered by the optimized test case x_i is first compared with the other remaining paths. Each remaining path is assigned a weight value based on p_{x_i} with its path encoding. The more identical nodes in the path code of a remaining path compared with p_{x_i}, the higher the weight value of that path. A remaining path is selected as the target using the roulette method based on the weight value of each remaining path. In the example program shown in Fig. 3.6, suppose $x_{old} = (1,10,100)$, whose corresponding path encoding is "No, No, No." The corresponding target path encoding p_{target} is "No, Yes, No." By comparing the two path encodings, it is found that the second branch node takes a different direction. The second column of the relationship matrix $(0,45,0)$ will be extracted. Since the relationship matrix shows that only the second dimension affects the direction of this branch, a dynamic scatter search is next performed on the second dimension of x_{old}.

The test-case-path relationship matrix collects the correlations between test case variables and path nodes to find the homogeneous subspace to cover the target path. The homogeneous subspace denotes the search space consisting of the test case coding dimensions related to the target path. Through the information-guided algorithm in the relationship

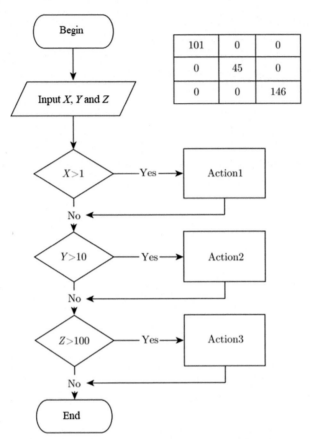

Figure 3.6 Flowchart of an example program and corresponding test-case-path relationship matrix.

matrix, more fitness evaluations are allocated to this subspace. Thus, it reduces the test case overhead of the algorithm and improves the search efficiency of the DE algorithm.

3.1.2.3 Experimental results and discussion
The experiments conducted in this section are based on the benchmark programs selected from the iFogSim toolkit [1]. The parameters of the benchmarks are shown in Fig. 3.7. A total of two sets of experiments were conducted.

The first experiment is the comparison results between RP-DE and DE under the mathematical model [2]. The results show that the relationship

Program	Loc	Dim	Description	Path
transmit	30	3	transmit sensor data to another sensor	2
send	47	2	send packaged tuple from sensor to fog-device or from one fog-device to another fog-device	9
processEvent	67	7	fog-device process events	9
executeTuple	41	7	use tuple processing logic to update device power consumption	5
checkCloudletCompletion	43	5	be called on fog-device on completion of execution of the tuple	6
getResultantTuple	73	8	return the processed tuple to applications	7

Figure 3.7 Detail of benchmark programs.

Function	RP-DE		DE	
ID	ave.c	ave.T(std.T)	ave.c	ave.T(std.T)
1	100%	4.20E+03(1.32E+03)	100%	6.13E+03(1.16E+03)+
2	66.7%	3.00E+05(0)	57.6%	3.00E+05(0)=
3	100%	1.41E+04(6.25E+03)	79.9%	3.00E+05(0)+
4	100%	9.40E+04(4.79E+04)	80.7%	2.92E+05(1.55E+04)+
5	100%	9.81E+03(2.94E+03)	98.9%	4.26E+04(4.42E+04)+
6	97.5%	1.57E+05(6.75E+04)	35.6%	3.00E+05(0)+
+/=/−				5/1/0

Figure 3.8 Comparison results of the first experiment.

Function	RP-DE		IGA		ABC		PSO	
ID	ave.c	ave.T(std.T)	ave.c	ave.T(std.T)	ave.c	ave.T(std.T)	ave.c	ave.T(std.T)
1	100%	4.20E+03(1.32E+03)	100%	1.00E+04(3.99E+03)+	53.3%	2.96E+05(2.06E+04)+	50%	3.00E+05(0)+
2	66.7%	3.00E+05(0)	43.3%	3.00E+05(0)=	34.8%	3.00E+05(0)=	44%	3.00E+05(0)=
3	100%	1.41E+04(6.25E+03)	70.4%	3.00E+05(0)+	83.3%	2.95E+05(2.21E+05)+	79.2%	3.00E+05(0)+
4	100%	9.40E+04(4.79E+04)	86%	2.30E+05(1.02E+04)+	40.7%	3.00E+05(0)+	40.7%	3.00E+05(0)+
5	100%	9.81E+03(2.94E+03)	98.9%	5.57E+04(5.19E+04)+	67.1%	3.00E+05(0)+	68.3%	3.00E+05(0)+
6	97.5%	1.57E+05(6.75E+04)	14%	3.00E+05(0)+	14.9%	3.00E+05(0)+	14.5%	3.00E+05(0)+
+/=/−				5/1/0		5/1/0		5/1/0

(a)

Function	RP-IGA		RP-ABC		RP-PSO		Random	
ID	ave.c	ave.T(std.T)	ave.c	ave.T(std.T)	ave.c	ave.T(std.T)	ave.c	ave.T(std.T)
1	100%	3.97E+04(1.59E+04)+	56.7%	2.83E+05(3.04E+04)+	58.3%	2.75E+05(4.40E+04)+	51.7%	3.00E+05(0)+
2	66.7%	3.00E+05(0)=	66.7%	3.00E+05(0)=	66.7%	3.00E+05(0)=	35.6%	3.00E+05(0)=
3	100%	1.47E+04(1.43E+04)+	100%	2.44E+03(1.55E+03)−	100%	6.05E+03(5.46E+03)−	88.5%	3.00E+05(0)+
4	58.7%	2.64E+05(6.04E+04)+	44%	3.00E+05(0)+	42.7%	3.00E+05(0)+	40.7%	3.00E+05(0)+
5	96.7%	1.02E+05(7.45E+04)+	75.6%	2.91E+05(1.86E+04)+	76.2%	3.00E+05(0)+	67.1%	3.00E+05(0)+
6	43.9%	2.77E+05(3.84E+04)+	16.8%	3.00E+05(0)+	15.9%	3.00E+05(0)+	14.5%	3.00E+05(0)+
+/=/−		4/2/0		4/1/1		4/1/1		5/1/0

Figure 3.9 Comparison results of the second experiment.

matrix can allocate more fitness evaluations of the DE algorithm to the low-dimensional Euclidean space which is the search space of relevant dimensions. Thus, it improves the ability of the DE algorithm for ATCG-PC in fog computing (Fig. 3.8).

The second experiment in Fig. 3.9 is the comparison between RP-DE with improved genetic algorithm (IGA), artificial bee colony (ABC), particle swarm optimization (PSO), and other algorithm variants [2]. The experimental results show that the relationship matrix improves the efficiency of the IGA, ABC, and PSO algorithms in solving the ATCG–PC problem in the iFogSim tool by allocating more fitness evaluations of the algorithms to search the low-dimensional Euclidean space.

3.1.3 Summary

This section introduces a test-case-path relationship matrix optimization strategy for the ATCG problem in the iFogSim toolkit. It is combined with evolutionary algorithms such as the DE algorithm. The algorithm introduces a test-case-path relationship matrix that records the correlation between test case dimensions and path branches. This allows the algorithm to update the matrix values corresponding to the position of the dimension when a new path is covered by changing the dimension value during the search process. In the subsequent update process, DE can use the relationship matrix to find out the relevant dimensional information and the target-related subspace. It approximately reduces the size of the search space and thus accelerates the convergence of the algorithm.

References: [2,3]

Related codes: http://www2.scut.edu.cn/huanghan/fblw/list.htm.

3.2 Effective natural language processing programs testing by random heuristic algorithm and scatter search strategy

3.2.1 Overview of research progress

As a theory-driven computational intelligence technology, natural language processing (NLP) currently has important application prospects in technology fields such as automatic web recommendations, social sentiment analysis, and chatbots. As shown in Fig. 3.10, NLP programs are widely used in various areas of society. They are mainly used to systematically analyze, understand, and extract information from text data in an intelligent and efficient manner. However, it is necessary to develop an efficient software testing tool to ensure the reliability of NLP programs and to reduce the cost of testing.

To minimize the test case overhead of the algorithm, ATCG-PC of NLP programs is modeled as a single objective optimization problem. There are two difficulties in solving this problem. Firstly, there are many string input variables in the NLP program. However, some paths can be covered only when specific values of strings are input test cases, which causes little effect in traditional optimization algorithms for solving path coverage test cases. Besides, there may be many test cases covering the same path in ATCG-PC. This can lead to a large amount of redundant

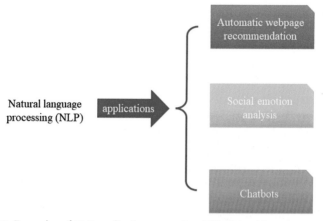

Figure 3.10 Examples of NLP application scenarios. *NLP*, Natural language processing.

overhead during optimization algorithm processing. To address the above problems, a search-based algorithm based on the scatter search strategy (SA-SS) [4] is introduced by the Intelligent Algorithm Research Center, School of Software Engineering, South China University of Technology. This strategy mainly optimizes the population and performs iterative updates repeatedly after combining random heuristic algorithms such as DE and PSO to solve the ATCG-PC problem of NLP programs.

3.2.2 Scientific principles
3.2.2.1 Problem description
This section presents a mathematical model for the ATCG based on path coverage of NLP programs. The goal of this mathematical model is to minimize the fitness evaluations within a certain target coverage, where $X = \{x_1, x_2, ..., x_N\}$ denotes the set of all test cases in the test case encoding space. $P = \{p_1, p_2, ..., p_L\}$ denotes the set of feasible paths of test functions. The ATCG-PC problem can be modeled as follows.

Minimize m (fitness evaluations)

s.t.

$$\theta_{ij} = \begin{cases} 1, & \text{test case } X_i \text{ covers path } p_j, \\ 0, & \text{otherwise} \end{cases} \tag{3.7}$$

$$\sum_{j}^{L} \theta_{ij} = 1 \tag{3.8}$$

$$\sum_{j=1}^{L} \min\left\{1, \sum_{i=1}^{T} \theta_{ij}\right\} = L \qquad (3.9)$$

$$m \leq M \qquad (3.10)$$

$$x_{n_i} \in S_\theta;\ S_\theta \in X;\ i = 1, 2, \ldots, m;\ m = |S_\theta|;$$
$$j = 1, 2, \ldots, L; 1 \leq n_1 \leq n_2 \leq \ldots \leq n_m \leq N; \qquad (3.11)$$

where Constraints (3.7) and (3.8) define the intermediate variable θ_{ij}. θ_{ij} equals zero only if $fitness_j(x_{n_i}) = 0$ when the test case x_{n_i} covers the jth path p_j. One test case can only cover only one path. Constraints (3.9) and (3.10) define that all paths are covered, and that the fitness evaluation overhead of the algorithm is less than the predefined maximum M. Constraint (3.11) defines the set of test cases S_θ generated during the algorithm search. S_θ is a subset of the set X of all feasible test cases.

In addition to the difficulties of the ATCG-PC problem described in this section, there are two other assumptions for the design of the NLP program. One is the presence of a large number of string input variables in the NLP program. Some paths can be covered only when the input string is a specific value. Another is that an SA-SS can find test cases that match the rules of the input strings as soon as possible.

As shown in Fig. 3.11, the path encoded as "Yes, Yes" is only covered when the input string s is "Answerstring." Assuming that all characters take the value [0,255], the probability of a randomly generated string overwriting the path will be less than 8.28×10^{-28}.

3.2.2.2 Automated path coverage test case generation based on random heuristic algorithm with scatter search strategy

The process of SA-SS is shown in Fig. 3.12. This strategy first initializes the population, evaluates the population individuals, and updates the path coverage. The algorithm can update the population by using the update strategy of DE, PSO, and other algorithms to further evaluate the population individuals and update the path coverage. It can also update the whole population by SA-SS. Among them, the SA-SS is used to cover the target path with the following two steps. In the first step, individuals are optimized to generate target paths and the step length of scattering points is initialized. The second step starts to search all variable dimensions sequentially, and each scattering point is centered on the original optimal value, leaving the

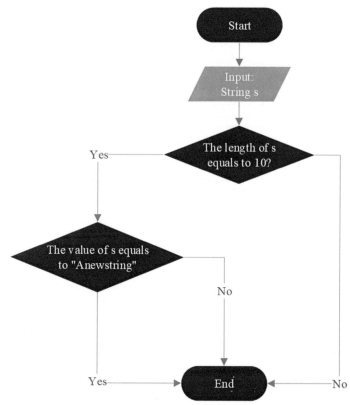

Figure 3.11 Flowchart of an NLP program. *NLP*, Natural language processing.

optimal individual, and updating the scattering step. This step is followed until the minimum search step is reached and all variables are traversed. Fig. 3.13 shows an example of optimizing a test case based on SA-SS.

Random heuristics such as DE, IGA, ABC, PSO, and competitive swarm optimization (CSO) are not efficient in solving ATCG-PC problems. In this work, four sets of simulation experiments are conducted to verify the effectiveness of the search strategy SA-SS by the strategy combining the above algorithms with the introduced strategy. These are four sets of experiments. (1) Comparison experiments of parameter s in SA-SS. (2) Comparison experiments of SA-SS that used differential evolution (DE-SS) with DE. (3) Comparison experiments of DE-SS with other algorithms (including IGA, ABC, PSO, CSO) combined with the SA-SS. (4) Comparison experiments of DE-SS with other algorithms (including IGA, ABC, PSO, CSO) combined with variable neighborhood search (VNS) algorithms.

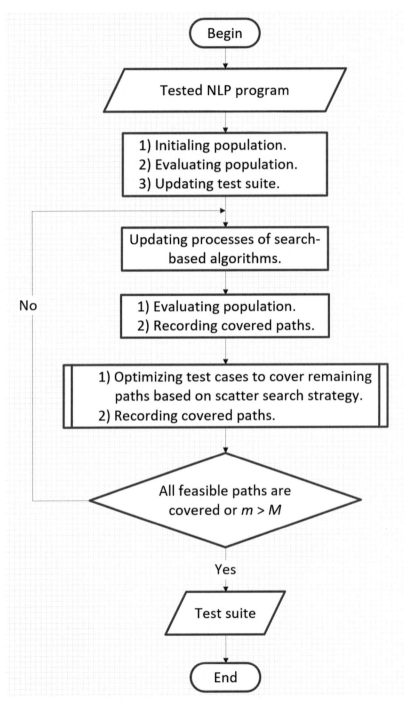

Figure 3.12 Flowchart of SA-SS. *SA-SS*, Scatter search strategy.

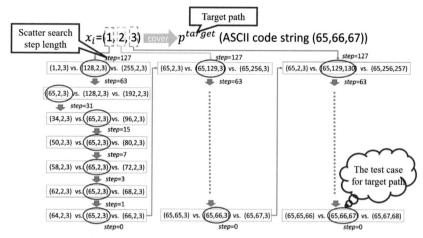

Figure 3.13 Diagram of solving ATCG-PC based on SA-SS. *ATCG-PC*, Automated test case generation for path coverage; *SA-SS*, scatter search strategy.

No.	Program	Loc	Dim	Description	Path
1	initFactory	86	7	to return the right type of token based on the options in the properties file and the type	48
2	cleanXmlAnnotator	21	6	to create a new object of cleanXmlAnnotator class	3
3	wordsToSentenceAnnotator	108	11	to transform a text of natural language to an annotator type	12
4	annotate	30	4	to turn the annotation into a sentence	3
5	nerClassifierCombiner	30	11	to create a new object of nerClassifierCombiner class	4
6	setTrueCaseText	37	6	to set the attribute of class trueCaseText	10

Figure 3.14 Detail of benchmark programs.

The subspace consists of multiple single-dimensional subspaces, and SA-SS can quickly find path coverage test cases for NLP programs by searching on this predefined subspace which can reduce the redundant fitness evaluation overhead. The subspace or subset of solutions satisfying the hypothesis is a set of multiple single-dimensional subspaces.

3.2.2.3 Experimental results and discussion

The benchmark programs in the experiments are selected from the NLP toolkit StandFord CoreNLP [5]. As shown in Fig. 3.14, Stanford CoreNLP is a collection of tools for processing natural language. It can get the basic form of different words (lexicality) and mark the structure of sentences based on phrases and grammatical dependencies to discover the relationships between entities, emotional color, and what people say. In addition, Stanford CoreNLP provides numerous tools for grammatical analysis. The application programming interfaces (APIs) of Stanford CoreNLP support most major (human) languages and work with most major programming languages. They are widely used for text analysis in all types of natural languages, providing the highest overall level of text analysis. Moreover, CoreNLP for natural language analysis also provides the basic building blocks for higher-level and domain-specific text understanding applications. It holds great significance in research and application areas involving NLP.

The first experiment focuses on analyzing the performance of parameter s in SA-SS [4]. DE is selected as the corresponding random heuristic algorithm, and the experimental comparison results are shown in Fig. 3.15. DE-SS algorithm performs best when the parameter s is taken as 2. However, the performance of DE-SS shows a gradual decrease as the setting value increases.

The comparison of DE and DE-SS is given in Fig. 3.15 [4]. As shown in Fig. 3.16, the SA-SS introduced in this work significantly reduces the test case overhead of the DE algorithm in solving the ATCG-PC problem. Moreover, DE-SS can achieve faster path coverage than DE within the same test case overhead.

The third experiment presents the results of comparing DE-SS with other related algorithms (including IGA, ABC, PSO, and CSO) [4]. As

Function ID	DE-SS-2		DE-SS-3	
	Ave.m(Std.m)	Rate	Ave.m(Std.m)	Rate
1	7.42E+03(1.32E+03)	100%	6.23E+04(6.21E+03)	100%
2	2.81E+02(2.59E+00)	100%	4.22E+02(1.03E+02)	100%
3	2.24E+03(4.42E+02)	100%	2.73E+03(6.39E+02)	100%
4	4.30E+02(1.42E+02)	100%	2.05E+04(7.01E+03)	100%
5	6.48E+02(1.92E+02)	100%	7.14E+02(2.39E+03)	100%
6	1.51E+03(2.35E+02)	100%	4.85E+03(4.23E+02)	100%
+/=/−			6/0/0	
Function ID	DE-SS-5		DE-SS-10	
	Ave.m(Std.m)	Rate	Ave.m(Std.m)	Rate
1	2.26E+05(8.73E+04)	46.7%	2.83E+05(3.62E+04)	20%
2	1.16E+04(7.83E+03)	100%	1.38E+04(9.59E+03)	100%
3	2.64E+04(1.98E+04)	100%	3.10E+04(1.44E+04)	100%
4	2.43E+04(5.78E+03)	100%	2.40E+04(6.69E+03)	100%
5	1.27E+05(8.19E+04)	93.3%	1.36E+05(7.39E+04)	90%
6	9.99E+04(4.68E+04)	100%	8.99E+04(2.80E+04)	100%
+/=/−	6/0/0		6/0/0	

Figure 3.15 Experimental results between DE-SS and DE. *DE*, Differential evolution; *DE-SS*, SA-SS that used differential evolution.

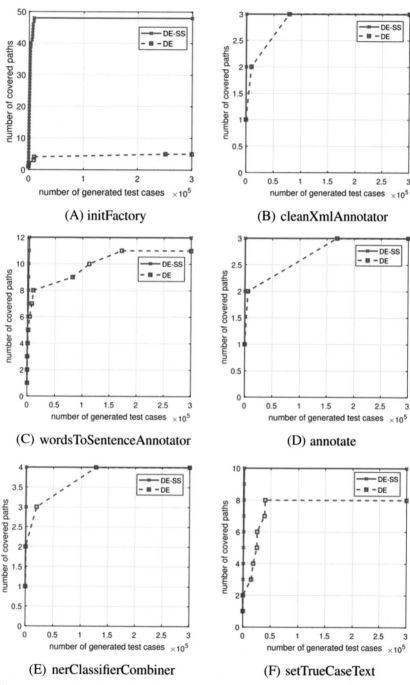

Figure 3.16 Path coverage number between DE-SS and DE. (A)–(F) correspond to the experimental results on the six benchmarks shown in Fig. 3.14, respectively. *DE*, Differential evolution; *DE-SS*, SA-SS that used differential evolution.

Function ID	DE-SS			IGA			IGA-SS		
	Ave.m(Std.m)	Rate	Time	Ave.m(Std.m)	Rate	Time	Ave.m(Std.m)	Rate	Time
1	7.14E+03(9.75E+02)	100%	7.40E00	3.00E+05(0)	0	3.64E+02	7.80E+03(1.60E+03)	100%	8.63E00
2	2.79E+02(3.79E+00)	100%	7.33E−01	2.90E+05(1.91E+04)	3.3%	3.11E+02	2.04E+02(2.55E+00)	100%	7.67E−01
3	2.21E+03(5.20E+02)	100%	3.33E00	2.91E+05(1.78E+04)	3.3%	3.91E+02	1.94E+04(1.45E+04)	100%	1.51E+01
4	4.26E+02(1.70E+02)	100%	9.33E−01	3.00E+05(0)	0	2.13E+02	2.91E+05(1.78E+04)	100%	1.27E00
5	5.82E+02(1.66E+02)	100%	1.63E00	1.26E+05(7.34E+04)	86.7%	1.02E+02	2.24E+03(2.94E+03)	100%	2.33E00
6	1.40E+03(1.65E+02)	100%	2.07E00	3.00E+05(0)	0	3.42E+02	1.13E+04(4.57E+03)	100%	9.53E00
+/=/−				6/0/0			2/3/1		

Function ID	DE-SS			ABC			ABC-SS		
	Ave.m(Std.m)	Rate	Time	Ave.m(Std.m)	Rate	Time	Ave.m(Std.m)	Rate	Time
1	7.14E+03(9.75E+02)	100%	7.40E00	3.00E+05(0)	0	1.69E+02	7.78E+03(1.14E+03)	100%	6.13E00
2	2.79E+02(3.79E+00)	100%	7.33E−01	3.00E+05(0)	0	1.33E+02	2.88E+02(7.52E+00)	100%	1.2E00
3	2.21E+03(5.20E+02)	100%	3.33E00	3.00E+05(0)	0	1.21E+02	2.44E+03(6.03E+02)	100%	3.3E00
4	4.26E+02(1.70E+02)	100%	9.33E−01	3.00E+05(0)	0	9.14E+01	4.73E+02(3.01E+02)	100%	1.03E00
5	5.82E+02(1.66E+02)	100%	1.63E00	3.00E+05(0)	0	1.50E+02	8.86E+02(3.99E+02)	100%	1.3E00
6	1.40E+03(1.65E+02)	100%	2.07E00	3.00E+05(0)	0	1.36E+02	1.85E+03(4.57E+02)	100%	2.87E00
+/=/−				6/0/0			3/3/0		

Function ID	DE-SS			PSO			PSO-SS		
	Ave.m(Std.m)	Rate	Time	Ave.m(Std.m)	Rate	Time	Ave.m(Std.m)	Rate	Time
1	7.14E+03(9.75E+02)	100%	7.40E00	3.00E+05(0)	0	2.60E+02	8.54E+03(2.27E+03)	100%	6.27E00
2	2.79E+02(3.79E+00)	100%	7.33E−01	3.00E+05(0)	0	2.27E+02	2.30E+02(3.79E+00)	100%	1.07E00
3	2.21E+03(5.20E+02)	100%	3.33E00	3.00E+05(0)	0	3.33E+02	3.34E+03(1.30E+03)	100%	5.43E00
4	4.26E+02(1.70E+02)	100%	9.33E−01	3.00E+05(0)	0	1.48E+02	4.99E+02(2.56E+02)	100%	1.03E00
5	5.82E+02(1.66E+02)	100%	1.63E00	3.00E+05(0)	0	2.45E+02	7.67E+02(3.83E+02)	100%	1.43E00
6	1.40E+03(1.65E+02)	100%	2.07E00	3.00E+05(0)	0	2.14E+02	2.01E+03(5.31E+02)	100%	2.27E00
+/=/−				6/0/0			4/1/1		

Function ID	DE-SS			CSO			CSO-SS		
	Ave.m(Std.m)	Rate	Time	Ave.m(Std.m)	Rate	Time	Ave.m(Std.m)	Rate	Time
1	7.14E+03(9.75E+02)	100%	7.40E00	3.00E+05(0)	0	2.36E+02	7.97E+03(1.68E+03)	100%	8.33E00
2	2.79E+02(3.79E+00)	100%	7.33E−01	2.70E+05(5.56E+04)	10%	1.82E+02	2.05E+02(3.41E+00)	100%	1.17E00
3	2.21E+03(5.20E+02)	100%	3.33E00	3.00E+05(0)	0	3.28E+02	2.52E+03(7.27E+02)	100%	5.07E00
4	4.26E+02(1.70E+02)	100%	9.33E−01	3.00E+05(0)	0	1.51E+02	4.10E+02(2.12E+02)	100%	9.00E−01
5	5.83E+02(1.66E+02)	100%	1.63E00	3.00E+05(0)	0	2.56E+02	5.62E+02(2.13E+02)	100%	1.17E00
6	1.40E+03(1.65E+02)	100%	2.07E00	3.00E+05(0)	0	2.14E+02	1.50E+03(3.63E+02)	100%	2.23E00
+/=/−				6/0/0			1/4/1		

Figure 3.17 Comparison between DE-SS and other compared algorithms. *DE-SS, SA-SS* that used differential evolution.

shown in Fig. 3.17, the SA-SS proposed in this research work can significantly improve the efficiency of algorithms such as IGA, ABC, PSO, and CSO in automatically generating NLP program path coverage test cases. The main aspects include the reduction of the number of test cases and the overhead of running time.

The fourth experiment presents the comparative analysis of DE-SS and VNS domain search strategies [4]. As shown in Fig. 3.18, the SA-SS introduced in this work has significant advantages over the traditional neighborhood search strategy such as VNS in solving the ATCG-PC problem.

In this study, we investigate how to solve the ATCG based on path coverage for NLP programs. It is a common and feasible idea to use random heuristics such as DE and PSO to solve a complex constrained optimization NP-hard problem. However, ATCG-PC for NLP programs has the difficulty that a large number of paths need to be covered by a specific value of the input string, which makes it difficult for traditional algorithms to meet the efficiency requirements when solving this problem. Therefore, this study performs unit tests on CoreNLP, a widely used NLP toolkit. A random heuristic algorithm based on the SA-SS is designed to

Function ID	DE-SS		DE-VNS		ABC-VNS	
	Ave.m(Std.m)	Rate	Ave.m(Std.m)	Rate	Ave.m(Std.m)	Rate
1	7.14E+03(9.75E+02)	100%	3.00E+05(0)	0	3.00E+05(0)	0
2	2.79E+02(3.79E+00)	100%	1.29E+03(0)	100%	1.34E+03(1.02E+02)	100%
3	2.21E+03(5.20E+02)	100%	3.00E+05(0)	0	3.00E+05(0)	0
4	4.26E+02(1.70E+02)	100%	8.82E+04(0)	0	7.07E+04(2.89E+04)	100%
5	5.82E+02(1.66E+02)	100%	2.86E+03(1.24E+03)	100%	4.36E+03(2.18E+03)	100%
6	1.40E+03(1.65E+02)	100%	9.98E+03(2.80E+03)	100%	1.25E+04(5.79E+03)	100%
+/=/−			6/0/0		6/0/0	

Function ID	IGA-VNS		PSO-VNS		CSO-VNS	
	Ave.m(Std.m)	Rate	Ave.m(Std.m)	Rate	Ave.m(Std.m)	Rate
1	3.00E+05(0)	0	3.00E+05(0)	0	3.00E+05(0)	0
2	1.26E+03(1.01E+02)	100%	1.29E+03(1.02E+02)	100%	2.95E+05(9.13E+03)	16.7%
3	3.00E+05(0)	0	3.00E+05(0)	0	3.00E+05(0)	0
4	8.81E+04(4.14E−01)	100%	3.00E+05(0)	0	8.81E+04(8.24E00)	100%
5	9.66E+03(1.31E+04)	100%	4.64E+03(3.45E+03)	100%	3.95E+03(2.32E+03)	100%
6	6.62E+04(6.99E+04)	96.7%	1.34E+04(7.26E+03)	100%	8.90E+04(2.16E+03)	100%
+/=/−	6/0/0		6/0/0		6/0/0	

Figure 3.18 Comparison between DE-SS and VNS. *DE-SS*, SA-SS that used differential evolution; *VNS*, Variable neighborhood search.

automatically generate path coverage test cases for CoreNLP programs. This work provides design ideas for algorithms to automatically generate paths to cover test cases for NLP programs of the same type. It enables researchers to design better and more efficient algorithms to test NLP programs based on this strategy.

In addition, the SA-SS introduced in this study can significantly reduce the test case overhead of the algorithm. It also improves the performance of the algorithm when automatically generating path coverage test cases for NLP programs or other programs that require specific input values to cover specific paths. By searching on the local manifolds that are equivalent to the target path, the algorithm's search range is reduced and the solution results are significantly improved. This idea of equating the search target to local manifolds is expected to provide more application examples and the theoretical foundation for machine learning, data mining, and other research fields.

3.2.3 Summary

In this chapter, an SA-SS is proposed for the characteristics of path distribution of NLP programs. This strategy can be combined with different heuristic algorithms such as the DE to efficiently solve the ATCG-PC problem. There are many paths in NLP programs that need to be covered when the input string is a specific value. It would be computationally overwhelming to search a decision space consisting of multiple string dimensions at the same time. In this section, an SA-SS is introduced based on the property that its subproblem dimensions do not affect each other. The strategy can generate path coverage test cases for NLP programs by dynamically searching the optimal solution of each dimension in the order of string variables. The technique has strong application value as an ATCG-PC problem in cutting-edge artificial intelligence systems.

References: [2−6]

Related Codes: http://www2.scut.edu.cn/huanghan/fblw/list.htm.

References

[1] Gupta H, Vahid Dastjerdi A, Ghosh SK, et al. iFogSim: a toolkit for modeling and simulation of resource management techniques in the Internet of Things, Edge and Fog computing environments. Software: Practice and Experience 2017;47(9):1275−96.

[2] Huang H, Liu F, Yang Z, et al. Automated test case generation based on differential evolution with relationship matrix for IFOGSIM toolkit. IEEE Transactions on Industrial Informatics 2018;14(11):5005−16.

[3] Huang H, Liu F, Zhuo X, et al. Differential evolution based on self-adaptive fitness function for automated test case generation. IEEE Computational Intelligence Magazine 2017;12(2):46−55.

[4] Liu F, Huang H, Yang Z, et al. Search-based algorithm with scatter search strategy for automated test case generation of NLP toolkit. IEEE Transactions on Emerging Topics in Computational Intelligence 2019;5(3):491−503.

[5] Manning CD, Surdeanu M, Bauer J, et al. The Stanford CoreNLP natural language processing toolkit. In Proceedings of 52nd annual meeting of the association for computational linguistics: system demonstrations, 2014;55−60.

[6] Liu F, Huang H, Li X, et al. Automated test data generation based on particle swarm optimization with convergence speed controller. CAAI Transactions on Intelligence Technology 2017;2(2):73−9.

CHAPTER 4

Application of multiobjective optimization intelligence algorithms

This chapter focuses on the application of intelligence algorithms in the field of multiobjective optimization, mainly including the application of intelligence algorithms to solve many-objective optimization problems (Section 4.1) and software product configuration problems (Section 4.2).

4.1 Many-objective evolutionary algorithm based on pareto-adaptive reference points

4.1.1 Overview of research progress

With the development of science and technology, multiobjective optimization problems (MaOPs) have been widely existed in our life. An MaOP refers to a problem with at least four optimization objectives, and objectives are often in conflict with each other. For example, production operators often want to get the maximum benefit with the minimum cost; when people purchase a car, they will consider the performance and comfort of the car in addition to the price. In general, as the number of objectives increases, the difficulty of solving the multiobjective optimization problems will also increase gradually.

More and more multiobjective evolutionary algorithms are proposed for solving different multiobjective problems. However, with the dramatic increase of data volume, it becomes a fundamental challenge that how to deal with the growth of objectives from a huge amount of information. For example, in the software product configuration problem, there are about 8.1×10^{17} software products without considering constraints when the dimension of the features is 25. In this case, how to reduce the solution time should be set as an objective. Therefore, how to design efficient multiobjective evolutionary algorithms is one of the current research hotspots.

Intelligent Algorithms
DOI: https://doi.org/10.1016/B978-0-443-21758-6.00004-8

In recent studies, the selection of suitable reference points (such as ideal points or nadir points) for different shapes of Pareto fronts (PF) is effective to address the above challenge. However, dominance resistant solution (DRS) is easy to generate by Pareto-dominance relationship-based algorithms, but they are difficult to detect and eliminate in iterations, which reduces the convergence speed of the algorithm.

To further address the above issues, a MaOEA with Pareto-adaptive reference points (PaRP/EA) [1] is introduced in this section. Specifically, the reference point at each generation is adaptively updated according to the shape of the PF, which is estimated by a ratio of Euclidean distances in the current population. In addition, an efficient method for detecting and eliminating DRS will be presented.

4.1.2 Scientific principles
4.1.2.1 General framework
The framework of PaRP/EA is shown in Algorithm 4.1. First, the initial population consists of solutions that are randomly generated in the whole decision space Ω. Two individuals are randomly selected as parents used to create offspring, as in NSGA-III [2] and VaEA [3]. In each iteration, the child solutions are generated by applying variation operations (i.e., crossover and mutation) to the parents. After repeating the above selection and variation operations $N/2$ times, an offspring population consisting of N new solutions can be obtained. Finally, the new population for the next generation is constructed by selecting N solutions from Q, the union of P and P'. The above steps are repeated until the termination criterion is met.

ALGORITHM 4.1 Framework of PaRP/EA.

Input: N (population size)
Output: The final population P.
1: $P \leftarrow initialization(N)$
2: **while** the termination criterion is not fulfilled **do**
3: $\quad P' \leftarrow variation(P)$ // generate an offspring population
4: $\quad Q \leftarrow P \cup P'$
5: $\quad P \leftarrow environmental\ Selection(Q)$ //environmental selection
6: **end while**
7: **return** P

4.1.2.2 Environmental selection

The pseudo-code of the environmental selection is given in Algorithm 4.2. The objective vectors in the population Q need to be normalized to address the issue that MaOPs have some scaled objective functions. Line 2 presents the same approach as in NSGA-III [2] to normalize individuals in Q. A detailed description of the normalization approach can be found below [2].

4.1.2.2.1 Adaptive normalization

For each objective $i = 1, 2, \ldots, m$, the minimum point $z^{min} = \left(z_1^{min}, z_2^{min}, \ldots, z_m^{min}\right)$ of the population Q is constructed by identifying the minimum value z_i^{min}. Consequently, the objective value of each individual can be transformed as $f_i'(x) = f_i(x) - z_i^{min}$, and the minimum point of the transformed population becomes a zero vector. Next, the determination of the

---

ALGORITHM 4.2 $P \leftarrow$ environmental Selection (Q).

Input: Q (the union population)
Output: The new population P
1: $P \leftarrow \varnothing$
2: $Q \leftarrow normalization(Q)$
3: $\{S_{nd}, S_d\} \leftarrow classificationByDominance$ // S_{nd} and S_d are the set of nondominated and dominated solutions, respectively.
4: $q \leftarrow estimateShapes(S_{nd})$// Estimate the fronts' shape, e.g., being linear, concave or convex
5: Set the reference point $\mathbf{r}$ to $\mathbf{z}^{nad}$ if $q < 0.9$, or $\mathbf{z}^*$ otherwise.
6: **if** $|S_{nd}| > N$ **then**
7:      $\{S_+, S_-\} \leftarrow classificationByHypercube(S_{nd})$//$S_+$ and S_- are resp. the sets of solutions inside and outside the hypercube, which is bounded by $\mathbf{z}^*$ and $\mathbf{z}^{nad}$.
8:      **if** $|S_+| > N$ **then**
9:          $P \leftarrow selection(S_+, S_-)$// Select N solutions from S_+ one by one
10:     **else**
11:         Add all members in S_+ into P, and then fill the population with the first $N - |P|$ best individuals in S_- according to fitness values.
12:     **end if**
13: **else**
14: Add all members in S_{nd} into P, and then fill the population with the first $N - |P|$ best individual in S_d according to fitness values.
15: **end if**
16: **return** P

extreme point $e_i \in \mathbb{R}^m$ along the objective axis f_i involves identifying the solution $x^i \in Q$ that minimizes the Achievement Scalarizing Function (ASF).

$$ASF(x, w_i) = \max_{j=1}^{m} \frac{f_j(x) - z_j^{\min}}{w_{i,j}} = \max_{j=1}^{m} \frac{f_j'}{w_{i,j}}, \qquad (4.1)$$

where $w_i = (w_{i,1}, w_{i,2}, \ldots, w_{i,m})$ denotes the axis direction of the objective axis f_i, which satisfies that if i not equal to j then $w_{i,j} = 0$, otherwise $w_{i,j} = 1$. Notably, the value 0 is set by a small positive value of 10^{-6} [4]. That is, $e_i = f(x_i) = (f_1(x_i), \ldots, f_m(x_i))$. After m extreme solutions have been generated, they are used to compose an $m - D$ hyperplane. The intercepts $a_i(i = 1, 2, \ldots, m)$ in the $i - th$ objective axis are utilized to further normalize the objective functions as

$$f_i''(x) = \frac{f_i'(x)}{a_i}, i = 1, 2, \ldots, m \qquad (4.2)$$

Note that the intercept on each normalized objective axis is now at $f_i'' = 1$ [2]. The hyperplane constructed based on these intercepts will satisfy $\sum_{i=1}^{m} f_i'' = 1$. For simplicity, the normalized objective $f_i''(x)$ is still represented as $f_i'(x)$.

Fig. 4.1 shows a two-objective illustration of the adaptive normalization procedure. In Fig. 4.1A, the minimum point $z^{\min}$ is first worked out for the population Q, then it is transferred to the origin Fig. 4.1B. Subsequently, the extreme solution along each objective axis can be determined by computing and comparing ASF values. In Fig. 4.1B, suppose $A = (10^{-6}, 30)$ and $B = (0.01, 3)$, then $ASF(A, w_2) = \max\{1, 30\}$ and $ASF(B, w_2) = \max\{\frac{0.01}{10^{-6}}, \frac{3}{1}\} = 10^4$, where $w_2 = (10^{-6}, 1)$. Since $ASF(A, w_2) < ASF(B, w_2)$, A is the extreme solution along f_2. However, A may not be a DRS in this case. According to the reference [5], *"DRSs are extremely inferior to others in at least one objective,"* but *"solutions that dominate a DRS are scarcely found."* DRSs are rephrased as *"those solutions with an extremely poor value in at least one of the objectives but with (nearly) optimal values in the others"* in reference [6], or *"DRSs are extremely poor in at least one of the objectives, while being extremely good in some others"* in reference [7]. Although the above definition of DRSs may lack clarity from a mathematical perspective [5], a common characteristic among DRSs that we can find is the existence of at least one extremely good value and one extremely poor objective value.

For the second objective value (i.e., 30) in $A = (10^{-6}, 30)$, it may not be extremely poor. From this point of view, A may not be a DRS. However, if $A = (10^{-6}, 10^5)$, $ASF(B, w_2) = \max\{\frac{10^{-6}}{10^{-6}}, \frac{10^5}{1}\} = 10^5$ is larger than

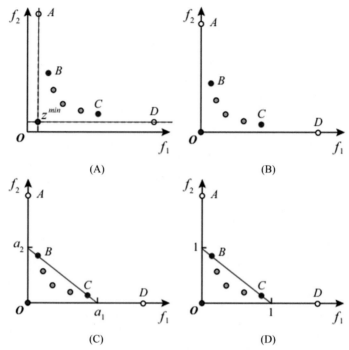

Figure 4.1 Illustration of the adaptive normalization. (A) The point z^{min} is worked out; (B) the z^{min} is transferred to 0; (C) extreme solutions B and C are determined; (D) intercepts a_1 and a_2 are transformed to 1.

$ASF(B, w_2) = 10^4$. In this case, B becomes the extreme solution, and A can be deemed as a DRS. Similarly, if $C = (3, 0.01)$ and $D = (10^5, 10^{-6})$, C is the extreme solution and D is a DRS along the $f_1 -$ axis. Assuming that both points B and C represent extreme solutions, as shown in Fig. 4.1C, it is possible to determine a straight line or, in the case of m >2, a plane or hyperplane based on these solutions. From Fig. 4.1C, the straight line is $\frac{f_1}{3.01} + \frac{f_2}{3.01} = 1$, indicating that the intercepts a1 and a2 are both 3.01. Finally, all individuals are normalized by Eq. (4.2) using these intercepts. As shown in Fig. 4.1D, the intercept on each normalized objective axis is at 1. Given the presence of DRSs, it is important to note that the normalized objectives may not fall within the range of [0, 1]. For example, A and D are normalized to $(3.3 \times 10^{-7}, 3.3 \times 10^4)$ and $(3.3 \times 10^4, 3.3 \times 10^{-7})$, respectively. These solutions need to be removed from the population. The reason for this can be found in Section 4.1.2. If both A and D are extreme solutions in Fig. 4.1C, it is noteworthy that the normalized objective values fall within the range of $0-1$.

The search process is guided by the shape of the PF. First, the mixed population Q is classified by using Pareto-dominance (Line 3 in Algorithm 4.2), i.e., $Q = S_{nd} \cup S_d$, where S_{nd} and S_d are the sets of non-dominated solutions and dominated solutions, respectively. Subsequently, solutions within S_{nd} are employed to estimate the shape of the PF (Line 4 in Algorithm 4.2). Specifically, the m nearest solutions in S_{nd} to the $m - D$ vector $v = (1, 1, \ldots, 1)$ are identified according to the angle between v and each of the solutions. Then the ratio

$$q = \frac{\bar{d}}{d^\perp} \tag{4.3}$$

can be used to determine the shape of the PF, where $\bar{d}$ and $d^\perp$ denote the average Euclidean distance from the m nearest solutions to the origin O and the Euclidean distance from O to the hyperplane $\sum_{i=1}^m f_i = 1$, respectively. Notably, $d^\perp$ is obtained by

$$d^\perp = \frac{|-1|}{\sqrt{m}} = \frac{1}{\sqrt{m}}$$

Eq. (4.3) can be rewritten as

$$q = \bar{d} . \sqrt{m} \tag{4.4}$$

4.1.2.2.2 Estimation of shapes and update of the reference point

An illustration is provided in Fig. 4.2. Fig. 4.2A shows an example of estimating the PF shape in a 2-D objective space. Since $\langle \vec{OC}, v \rangle$ and $\langle \vec{OD}, v \rangle$ are smaller than other angles, the two nearest solutions of the vector $v = (1, 1)$ are C and D. In this chapter, $\langle \cdot, \cdot \rangle$ denotes the angle between two different vectors. Therefore, according to Eq. (4.4), $q = \frac{\|\vec{OC}\| + \|\vec{OD}\|}{2} . \sqrt{2}$, where $\|\cdot\|$ is the L2-norm of a vector.

Fig. 4.2A shows a convex PF. In this case, $d^\perp$ is significantly larger than $\bar{d}$. Therefore, the ratio of $\bar{d}$ to $d^\perp$ (Eq. 4.3), q is obviously smaller than 1. Notably, a value of q smaller than 1 may indicate a convex PF. Furthermore, when q is notably greater than 1, the PF can be considered concave. Conversely, if q is very close to 1, the PF is likely to exhibit linearity. Due to the shape is estimated, we can use an interval to roughly judge whether the PF is linear or not. Therefore, the shape of the PF can be estimated as follows:

- The shape is convex, if q is less than 0.9;

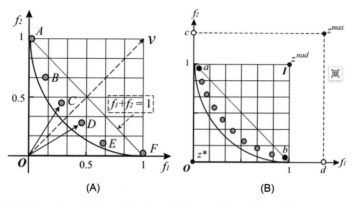

Figure 4.2 Illustration of the estimation of shapes (A) and the reference points (B).

- The shape is linear, if q is greater than or equal to 0.9 while less than or equal to 1.1;
- The shape is concave, if q is greater than 1.1.

Based on the estimated shape of the PF, the reference point $r = (r_1, r_2, \ldots, r_m)$ is dynamically updated at each generation (Line 5 in Algorithm 4.2). The reference point plays a crucial role in calculating the fitness values of solutions and the angles among them, significantly influencing the algorithm's performance. When the PF is convex, **r** is set to be the nadir point $z^{nad} = \left(z_1^{nad}, z_2^{nad}, \ldots, z_m^{nad}\right)$. Otherwise, **r** is set to be the ideal point $z* = \left(z_1^*, z_2^*, \ldots, z_m^*\right)$. As shown in Fig. 4.2B, z^{nad} and z^* are defined as the points $I = (1, 1, \ldots, 1)$ and $O = (0, 0, \ldots, 0)$, respectively. Note that the nadir point is not defined as the maximum point $z^{max} = \left(z_1^{max}, z_2^{max}, \ldots, z_m^{max}\right)$ from Fig. 4.2B, where z_i^{max} is the worst value for the i-th objective ($i = 1, 2, \ldots, m$) in the current population. The reason will be provided in Section 4.1.2.

4.1.2.2.3 Fitness assignment

The fitness value mainly measures the convergence of a solution. According to reference [8], the fitness of a solution should be assessed based on the shape of the PF.

- If the estimated shape is linear, the fitness is calculated by Eq. (4.5):

$$c(x) = \sum_{i=1}^{m} f_i(x) \tag{4.5}$$

This is the sum of all the objectives. Intuitively, it is well suited for problems whose PFs are linear.

- Given that the estimated shape is concave, the function c(x) is then determined by

$$c(x) = \sqrt{\sum_{i=1}^{m} (f_i(x) - r_i)^2} \qquad (4.6)$$

where $r_i = z_i^*$. It calculates the Euclidean distance from the current solution to the reference point (i.e. z*), and is widely used in algorithm designs [3,8,9].

- If the estimated shape is convex, c(x) is calculated by Eq. (4.7)

$$c(x) = \frac{1}{\sqrt{\sum_{i=1}^{m} (f_i(x) - r_i)^2}} \qquad (4.7)$$

where $r_i = r_i^{nad}$. In this case, the objective is to maximize the Euclidean distance between the current solution and the reference point (i.e. z^{nad}), thereby pushing the solutions away from the nadir point as far as possible. Since the fitness values are to be minimized, we calculate the reciprocal of the Euclidean distance to the nadir point. All the aforementioned fitness estimators are computationally efficient and easily obtainable. The main reason for choosing these fitness estimators is that they are beneficial for problems whose PFs fit the contour lines formed by these estimators [8]. Further detailed discussions on this aspect can be found in [8].

4.1.2.2.4 Classification by a hypercube

As shown in Line 6 of Algorithm 4.2, the procedure checks whether the size of S_{nd} is larger than N or not. If $|S_{nd}| \leq N$, we definitely add all the members in S_{nd} to P, and then fill the population with the first $N - |P|$ elite individuals in S_- according to fitness values (see Lines 13−15 in Algorithm 4.2). If $|S_{nd}| > N$, which is often the case for problems with a large number of objectives, the solutions in S_{nd} are further divided into two sets based on a hypercube, which is bound by $z*$ and z^{nad} (Line 7 in Algorithm 4.2). The two sets are S_+ and S_- that contain solutions inside and outside the hypercube, respectively. For example, as shown in Fig. 4.2B, the set S_- has 2 solutions, i.e., c and d, while S_+ contains all the remaining 10 solutions.

In fact, the classification by the hypercube provides a feasible way to deal with DRSs [5]. Suppose $a = (0.01, 0.90)$, $b = (0.90, 0.01)$, $c = (0.0, 10^7)$, and $d = (10^7, 0.0)$ in Fig. 4.2B. Since c and d are nondominated with other solutions, the classification by Pareto-dominance (Line 3 in Algorithm 4.2) cannot eliminate them. However, the classification by the hypercube can successfully eliminate them. According to Eq. (4.1),

$$ASF(b, w_1) = \max\left\{\frac{0.90}{1}, \frac{0.01}{10^{-6}}\right\} = 10^4 \quad \text{and} \quad ASF(d, w_1) = \max\left\{\frac{10^7}{1}, \frac{0.0}{10^{-6}}\right\} = 10^7,$$

where $w_1 = (1, 0)$. As per Section 4.1.2, since ASF (b, $\mathbf{w}_1$) $<$ ASF (d, $\mathbf{w}_1$), the extreme solution in the objective axis f_1 is b, rather than d. Similarly, in the axis f_2, a is recognized as the extreme solution, not c. Hence, the hyperplane [the straight line in Fig. 4.2B] is constructed based on a and b. Given that the hypercube classification selectively retains solutions within the space $[0, 1] \times [0, 1]$, c and d are consequently discarded. This elimination proves beneficial for enhancing convergence, particularly in scenarios where DRSs are typically situated far from the true PF. This mechanism proves effective in filtering out DRSs during experiments involving problems characterized by a substantial number of local optimal PFs.

4.1.2.2.5 Select solutions one by one

After classification, a selection procedure will be executed to choose N solutions from S_+ one by one if the size of S_+ still exceeds N (Lines 8 and 9 in Algorithm 4.2). Otherwise, all the solutions in S_+ are added into P, and the new population is then filled with the first $N - |P|$ elite individuals in S_- according to fitness values (Line 11 in Algorithm 4.2). For detailed information on the selection procedure, refer to Algorithm 4.3. The selection routine first adds m extreme solutions into P, subsequently removes them from S_+. These extreme solutions adhere to the definition provided in adaptive normalization (see Section 4.1.2) and have been identified during that phase. Next, the remaining solutions are selected based on the remove_one_by_one procedure or the add_one_by_one procedure, which is determined according to the size of S_-. If $|S_-| > 0$, signifying the presence of poorly converged solutions. Consequently, the remove_one_by_one procedure which emphasizes both convergence and diversity of solutions is executed. Otherwise, the add_one_by_one procedure is executed that stresses particularly the diversity of solutions.

ALGORITHM 4.3 P → selection (S$_+$, S$_-$)

Input: S_+ and S_-

Output: A new population P

    1: Add m extreme solutions into P and remove them from S_+.

    2: **if** $|S_-| > 0$ **then**

    3: /*—— remove_one_by_one procedure ——*/

    4: **repeat**

    5:    Select a pair of solutions from S_+ with the minimum angle to each other, and then remove from S_+ the one with the worst fitness value.

    6:    **until** $|P| + |S_+| = N$

    7: Add the remaining solutions in S_+ into P.

    8: **else**

    9: /*—— add_one_by_one procedure ——*/

  10: **repeat**

  11:    Add into P the solution in S_+ that has the maximum angle to P.

  12:    **until** $|P| = N$

  13: **end if**

  14: **return** P

For the *remove_one_by_one* procedure, which is actually the elimination procedure used in our previous work [10] (Lines 4−7 in Algorithm 4.3), a pair of solutions are selected with the minimum angle among all solution pairs and then remove the worst one (measured by the fitness values) from S_+. The above operation is iteratively performed until the sum of the sizes of P and S_+ reaches N. Finally, all the remaining solutions in S_+ are amalgamated into P. As for the *add_one_by_one* procedure, it is the same as the *maximum-angle-first* principle introduced in our prior work [3]. During each iteration, the solution that is farthest is incorporated into the new population P. It should be noted that the "distance" between individuals is measured by angles rather than the Euclidean distance. For a comprehensive understanding of the *add_one_by_one* procedure, refer to the original work [3].

Both *remove_one_by_one* and *add_one_by_one* use angle to evaluate the density around solutions. In fact, it has already been well demonstrated that angles are effective in measuring distances between solutions, especially for MaOPs [3,8,10−14]. Therefore, we employ angle-based selection in PaRP/EA as well. This selection strategy is implemented by *remove_one_by_one* and *add_one_by_one*, which are derived from existing methods in [10] and [3], respectively. Different from the existing works where the reference point used for calculating angles is fixed to the ideal point, we dynamically update this point based on the estimated shape of the PF. Furthermore, only *remove_one_by_one* or *add_one_by_one* is used in [10] or [3]. In PaRP/EA, however, the above two procedures collaborate synergistically, making full use of each one's own advantages.

4.1.2.3 Experimental analysis

In this section, we illustrate the considerable effectiveness of the introduced PaRP/EA through computational experiments where PaRP/EA is compared with NSGA-III[1] [2], VaEA[2] [3], MOEA/D[3] [15], θ-DEA[4] [4], 1by1EA5 [8], GWASF-GA[6] [16], and MOEA/D-IPBI[7] [17]. The chosen algorithms have been shown to be effective in handling MaOPs, and they can be divided into two categories: the Pareto-dominance based algorithms (VaEA and 1by1EA) and the reference points/weight vectors based algorithms (NSGA-III, MOEA/D, θ-DEA, GWASF-GA, and MOEA/D-IPBI).

In this empirical study, we select 29 test problems, which are DTLZ1−DTLZ7 [18], ConvexDTLZ2 and ConvexDTLZ4 [2], DTLZ1^{-1} and DTLZ3^{-1} [19], WFG1−WFG9 [20], and WFG1^{-1}−WFG9^{-1} [19]. The selected problems have various types of PFs. For example, DTLZ2−4 and WFG4−9 have concave PFs, whereas DTLZ3^{-1}, ConvexDTLZ2, ConvexDTLZ4 and WFG4. 9^{-1} have convex PFs. The PFs of DTLZ5−7, WFG1−3, and WFG1−2^{-1} are complicated, mixed, discontinued, or degenerated. Note that the parameters in these test problems are configured in accordance with the recommendations provided in the original studies [2,18−20]. In the experiments, the well-known inverted generational distance (IGD) [15,21] and hypervolume (HV) [22] are used to evaluate the performance of the algorithms. Both IGD and HV are able to simultaneously measure the convergence and diversity of the obtained solution sets. When calculating IGD, we require a set of reference points sampled (nearly) uniformly on the true PFs. For DTLZ1−4, ConvexDTLZ2, ConvexDTLZ4, and WFG4−9, we use the same method as in [3] to obtain a set of reference points. For WFG3−9^{-1}, DTLZ1^{-1}, and DTLZ3^{-1}, we randomly generate 100,000 reference points on the true PF using the uniform distribution, following the practice in [19]. Since the specification of the uniform distribution over the true PF is difficult for WFG1−3, DTLZ5−7, and WFG1−2^{-1}, all nondominated solutions among all of the obtained solutions are used as reference points for these problems [19]. When calculating HV, according to [4,19], the objective values of points are first normalized to [0, 1] using the ideal and nadir points of the true PF (or the approximated PF). Then the reference point is specified as $(1.1, 1.1, \ldots, 1.1)$. This configuration aligns with the settings employed in both [19] and [4]. In addition, for $m \leq 10$, the recently introduced WFG algorithm [23] is used to calculate the exact HV. For $m = 15$, the HV is approximated by the Monte Carlo simulation method [24], where 10,000,000 sampling points are used to ensure accuracy.

4.1.2.4 Experimental settings

For each test problem, the number of objectives is 3, 5, 8, 10, and 15. All the algorithms are independently run 30 times in each test instance for each problem. General experimental settings are summarized as follows.

4.1.2.4.1 Population size and termination condition

The population size N in MOEA/D and MOEA/D-IPBI is set to be the number of weight vectors, which is 91, 210, 156, 275, and 135 for $m = 3$, 5, 8, 10, and 15, respectively. For PaRP/EA, NSGA-III, VaEA, θ-DEA, and 1by1EA, N is set to be the smallest multiple of four larger than the number of weight vectors, i.e., $N = 92$, 212, 156, 276, and 136 for $m = 3$, 5, 8, 10, and 15, respectively. As GWASF-GA utilizes a binary tournament selection procedure, the population size should be even. If the number of generated weight vectors is odd, a random one will be removed as suggested by its developers [16]. Therefore, N in GWASF-GA is 90, 210, 156, 274, and 134 for $m = 3$, 5, 8, 10, and 15, respectively. Note that the weight vectors are generated using Das and Dennis's systematic approach [25] ($m \leq 5$) and the two-layer weight vector generation method [2,26] ($m > 5$). To ensure a fair comparison, all the decomposition-based algorithms (i.e., NSGA-III, MOEA/D, θ-DEA, GWASF-GA, and MOEA/D-IPBI) use the same set of weight vectors. All the algorithms are terminated when the number of function evaluations (FEs) reaches 92,000 (i.e., $92 \times 1\ 000$), 265,000 (i.e., 212×1250), 234,000 (i.e., 156×1500), 552,000 (i.e., 276×2000), and 408,000 (i.e., $136 \times 3\ 000$) for 3-, 5-, 8-, 10-, and 15-objective test problems, respectively. Note that the first factor in the brackets represents the population size N for PaRP/EA, NSGA-III, whereas the second factor corresponds to the maximum number of generations. Since N in the algorithms may be different, it will be fair to compare them using the number of maximum function evaluations (*max_FEs*) instead of the maximum generations. Therefore, *max_FEs* is used as the termination condition in this section.

4.1.2.4.2 Parameter settings

In all the algorithms, the simulated binary crossover (SBX) and the polynomial mutation (PM) are used to generate offsprings. The parameters in the two genetic operators are set as follows: $p_c = 1.0$, $p_m = 1/n$, $\eta_c = 30$, and $\eta_m = 20$ [2,26], where p_c and p_m are the crossover and mutation probabilities, respectively; η_c and η_m are the distribution index for SBX and PM, respectively. We use standard toolkits to implement the peer algorithms, adhering to the parameter settings suggested in their original study. Specifically, the PBI is chosen as the decomposition approach in MOEA/D, where the penalty parameter θ is set to be 5 [15]. In MOEA/D, the neighborhood size T is configured as 20. Additionally, in θ-DEA, PBI is employed with θ set to 5, in accordance with the methodology outlined in [4]. According to the reference [8], the parameter k in 1by1EA is set to be 0.1 N, aiming to strike a balance between computational cost and accuracy in estimating density, and R is set to be 1 to ensure equilibrium convergence and distribution. Following the guidelines in [16], the parameter $\in i(i = 1, 2, \ldots, m)$ in GWASF-GA, which is used to define utopian objective vector, is set to be 1% of the difference between the corresponding nadir and ideal values. In MOEA/D-IPBI [17], the parameter θ is established at 0.1. Experimental results in [19] indicate that IPBI with $\theta = 0.1$ yields promising results across a majority of the test problems in the study. Similar to NSGA-III [2] and VaEA [3], the PaRP/EA does not use any algorithm parameters.

4.1.2.5 Experimental results on DTLZ test problems

In this section, the DTLZ test problems encompass not only the original DTLZ1−7 but also those adapted from them, i.e., ConvexDTLZ2, ConvexDTLZ4, DTLZ1^{-1}, and DTLZ3^{-1}. Therefore, we collected 11 DTLZ test problems in total. To facilitate the analysis process, we present the Wilcoxon's rank sum [27,28] test results in Table 4.1. As shown in the table, PaRP/EA significantly outperforms NSGA-III, VaEA, MOEA/D, θ-DEA, 1by1EA, GWASF-GA, and MOEA/D-IPBI in 43/55 = 78%, 38/55 = 69%, 51/55 = 93%, 44/55 = 80%, 36/55 = 65%, 52/55 = 95%, and 53/55 = 96% DTLZ test instances, respectively. From Table 4.1, the smallest and the largest percentages are 65% and 96%, respectively. Moreover, the highest percentage of the test instances where PaRP/EA exhibits significant inferiority compared to the peer algorithms is identified in the pairwise comparison between PaRP/EA and 1by1EA, being only 13/55 = 24%. With respect to the IGD results, PaRP/EA would be the most effective algorithm in handling DTLZ test problems. In terms of

Table 4.1 The proportion of DTLZ test instances where PaRP/EA is better than (•), worse than (○), and equal to (‡) each of the peer algorithms concerning inverted generational distance (IGD) and hypervolume (HV) metrics.

RaRP/EA v.s.		NSGA-III	VaEA	MOEA/D	θ-DEA	1by1EA	GWASF-GA	MOEA/D-IPBI
IGD	•	43/55	38/55	51/55	44/55	36/55	52/55	53/55
	○	10/55	11/55	3/55	10/55	13/55	2/55	2/55
	‡	2/55	6/55	1/55	1/55	6/55	1/55	0/55
HV	•	23/55	39/55	47/55	26/55	48/55	47/55	50/55
	○	23/55	4/55	4/55	21/55	3/55	1/55	0/55
	‡	9/55	12/55	4/55	8/55	4/55	7/55	5/55

HV results, as indicated in Table 4.1, PaRP/EA demonstrates competitiveness with NSGA-III and θ–DEA. Moreover, it significantly outperforms other algorithms in at least 39 out of 55 test instances. We can find that the largest ratio is 50/55 = 91%, which is detected in the pairwise comparison between PaRP/EA and MOEA/D-IPBI.

As the Pareto-adaptive reference points are used to fit the shapes of the fronts, we concentrate on the performance of the algorithms in handling problems characterized by convex fronts. Fig. 4.3 shows the final solutions of all the algorithms on the three-objective ConvexDTLZ2 problem. The solutions of PaRP/EA cover the whole convex PF more widely than those of other algorithms. The solutions produced by VaEA, MOEA/D, θ–DEA, and GWASF-GA are predominantly distributed in the central region of the PF. Since MOEA/D and θ–DEA use a set of systematically generated weight vectors, their solutions are distributed more regularly than those of VaEA. While GWASF-GA utilizes the same set of weight vectors, its ranking method places a greater emphasis on convergence rather than the diversity of solutions. As a result, the diversity of the final population of GWASF-GA may be inferior to that of MOEA/D, as also evident in [16]. Moreover, all the four algorithms fail to find solutions covering the boundaries well. According to Fig. 4.3F and H, the solutions of both 1by1EA and MOEA/D-IPBI encounter an obstacle in converging towards the true PF, which is clear from the large objective values found in the cartesian coordinate system. Although the solutions of NSGA-III from Fig. 4.3B are distributed regularly and can cover both the center and boundary of the PF, they actually do not cover the PF as widely as those obtained by PaRP/EA.

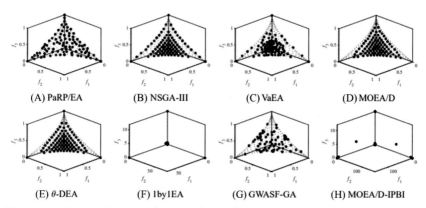

Figure 4.3 Final solutions obtained by each algorithm on the three-objective ConvexDTLZ2 test problem. (A) PaRP/EA; (B) NSGA-III; (C) VaEA; (D) MOEA/D; (E) θ-DEA; (F) 1by1EA; (G) GWASF-GA; (H) MOEA/D-IPBI.

4.1.2.6 Empirical findings for WFG and WFG^{-1} test problems

The results of the Wilcoxon's rank sum test on the WFG1−9 test problems are presented in Table 4.2. In terms of IGD performance, PaRP/EA demonstrates a significant enhancement compared to NSGA-III, VaEA, MOEA/D, θ-DEA, 1by1EA, GWASF-GA, and MOEA/D-IPBI in 28/45 (62%), 29/45 (64%), 36/45 (80%), 32/45 (71%), 38/45 (84%), 42/45 (93%), and 38/45 (84%) test instances, respectively. Regarding HV results, PaRP/EA shows comparable performance to NSGA-III and θ-DEA, outperforming all other algorithms significantly. The highest proportion of instances where PaRP/EA is outperformed by other algorithms is observed in the PaRP/EA vs. θ-DEA comparison, with only 9/45 (20%) instances. Table 4.3 summarizes the statistical test outcomes for the WFG^{-1} problems, considering both IGD and HV. Analysis of the IGD results reveals that PaRP/EA outperforms 1by1EA in all 45 test instances, achieving a 100% success rate, surpasses NSGA-III, MOEA/D, and θ-DEA in 44/45 instances (98%), and finally, surpasses both GWASF-GA and MOEA/D-IPBI in 39/45 instances (87%). Notably, VaEA emerges as the most formidable competitor to PaRP/EA. In terms of HV results, PaRP/EA outperforms other algorithms in at least 31 test instances (69%) and at most 38 test instances (84%). For all other algorithms, excluding GWASF-GA and MOEA/D-IPBI, PaRP/EA is only defeated by its competitors in a single test instance.

Fig. 4.4 illustrates the final solution sets for the three-objective WFG9^{-1}. Notably, PaRP/EA and MOEA/D-IPBI demonstrate superior performance in terms of the distribution of final solutions compared to other algorithms. While NSGA-III and θ-DEA exhibit solutions that lack sufficient distribution to cover the entire PF, 1by1EA displays a biased coverage towards one half of the PF. VaEA, MOEA/D, and GWASF-GA reveal the presence of clusters among their obtained solutions. Specifically, VaEA's solutions are predominantly situated in the central region of the PF, whereas MOEA/D and GWASF-GA exhibit clusters along the boundaries or at the corners. It is noteworthy that the solutions of MOEA/D-IPBI exhibit a more uniform distribution compared to PaRP/EA due to the utilization of systematically generated weight vectors in MOEA/D-IPBI.

In Fig. 4.5, the solution sets for the 15-objective WFG7^{-1} problem are depicted using parallel coordinates (indicated by the black lines). To highlight variations in algorithm performance across concave and convex problems, the solutions for the original WFG7 are concurrently presented in the same figure, marked by red lines. Examining Fig. 4.5D and G, both MOEA/D and GWASF-GA exhibit suboptimal performance on both WFG7 and WFG7^{-1},

Table 4.2 The proportion of Empirical Findings for WFG test problem where PaRP/EA is better than (•), worse than (○), and equal to (‡) each of the peer algorithms.

RaRP/EA v.s.		NSGA-III	VaEA	MOEA/D	θ-DEA	1by1EA	GWASF-GA	MOEA/D-IPBI
IGD	•	28/45	29/45	36/45	32/45	38/45	42/45	38/45
	○	11/45	3/45	5/45	9/45	1/45	0/45	4/45
	‡	6/45	13/45	4/45	4/45	6/45	3/45	3/45
HV	•	23/45	35/45	41/45	19/45	32/45	39/45	34/45
	○	23/45	1/45	1/45	9/45	6/45	4/45	4/45
	‡	9/45	9/45	3/45	17/45	7/45	2/45	7/45

HV, Hypervolume; *IGD*, inverted generational distance.

Table 4.3 The proportion of Empirical Findings for WFG^{-1} test problem where PaRP/EA is better than (•), worse than (○), and equal to (‡) each of the peer algorithms.

RaRP/EA v.s.		NSGA-III	VaEA	MOEA/D	θ-DEA	1by1EA	GWASF-GA	MOEA/D-IPBI
IGD	•	44/45	16/45	44/45	44/45	45/45	39/45	39/45
	○	1/45	8/45	0/45	0/45	0/45	3/45	6/45
	‡	0/45	21/45	1/45	1/45	0/45	3/45	0/45
HV	•	31/45	32/45	37/45	36/45	38/45	31/45	31/45
	○	1/45	1/45	1/45	1/45	1/45	2/45	6/45
	‡	13/45	12/45	7/45	8/45	6/45	12/45	8/45

HV, Hypervolume; *IGD*, inverted generational distance.

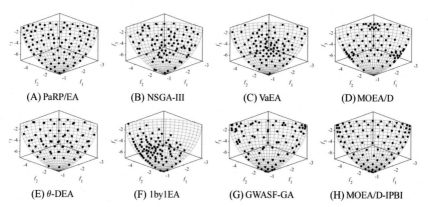

Figure 4.4 Final results of different algorithms on the three-objective WFG9^{-1} test problem. (A) PaRP/EA; (B) NSGA-III; (C) VaEA; (D) MOEA/D; (E) θ-DEA; (F) 1by1EA; (G) GWASF-GA; (H) MOEA/D-IPBI.

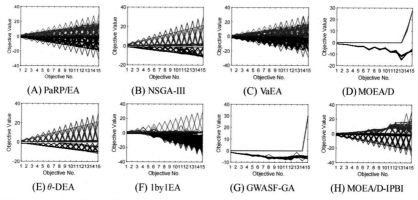

Figure 4.5 Final results of different algorithms on the 15-objective WFG7 and WFG7^{-1} test problems. (A) PaRP/EA; (B) NSGA-III; (C) VaEA; (D) MOEA/D; (E) θ-DEA; (F) 1by1EA; (G) GWASF-GA; (H) MOEA/D-IPBI.

displaying remarkably similar solutions in each case. Conversely, Fig. 4.5B and E reveal that NSGA-III and θ-DEA are better on the original WFG7 compared to its inverse counterpart, WFG7^{-1}. While MOEAD-IPBI struggles to provide diverse solutions for WFG7, it exhibits commendable performance on the WFG7^{-1} problem. These observations align with the conclusions in [19], emphasizing that the efficacy of decomposition–based algorithms is heavily influenced by the shapes of PFs. Algorithms employing systematically generated weight vectors, such as NSGA-III and θ-DEA, prove particularly adept at solving problems with triangular-shaped PFs (e.g., WFG4−9), as the distribution of weight vectors (or reference lines) aligns with the PF's shape. MOEAD-IPBI shifts all solutions from the nadir point to the PF, resulting in a distribution of weight vectors that aligns with the shape of inverted triangular PFs. This explains why MOEA/D-IPBI yields well-distributed solutions on WFG7^{-1}, characterized by an inverted triangular-like PF.

Let's assess the performance of the algorithms without employing weight vectors. Generally, there is no significant decline in performance when transitioning from WFG7 to WFG7^{-1}. Illustrated in Fig. 4.5F, the solutions produced by 1by1EA exhibit distinct advantages in their distribution across the two problems. On WFG7, the solutions effectively cover the PF in several first objectives but exhibit deficiencies, particularly in diversity, in others. Conversely, solutions on WFG7^{-1} demonstrate a broad coverage of the PF in several last objectives. Examining Fig. 4.5C, solutions generated by VaEA on both WFG7 and WFG7^{-1} display a well-distributed pattern; however, VaEA encounters challenges in obtaining lower bounds in certain objectives on the WFG7^{-1} problem (note that the PF satisfies $-2i - 1 \leq f_i \leq -1$ for $i = 1, 2, \ldots, 15$ [19]). As depicted in Fig. 4.5A, solutions derived from PaRP/EA exhibit widespread distribution for both WFG7 and WFG7^{-1}. The effectiveness of PaRP/EA, bolstered by Pareto-adaptive reference points, might render it immune to the concavity/convexity of the PF. In cases where the PF is deemed convex, the nadir point serves as the reference point; otherwise, the ideal point is used. As discussed elsewhere [16,17,19], the nadir point is deemed more suitable for a convex PF. Thus, with Pareto-adaptive reference points, PaRP/EA demonstrates adaptability to various PF shapes, elucidating its commendable performance across both concave and convex problems considered in this study.

4.1.3 Summary

This section explores the utilization of intelligent algorithms in multiobjective optimization, particularly in MaOPs. A straightforward and efficient approach is presented for approximating the shape of the PF and dynamically selecting reference points. Metrics for convergence and diversity are computed based on the adaptive selection of reference points. Furthermore, a practical method for identifying and eliminating DRSs is provided. Experimental results demonstrate that the adaptive reference points contribute to enhancing the algorithm's performance in handling PFs of diverse shapes, thereby increasing its flexibility. The introduced method for addressing DRSs effectively eliminates such solutions in a timely manner, consequently accelerating the algorithm's convergence and improving its overall performance. Additional materials include the following:

- Experimental data: https://ieeexplore.ieee.org/document/8682100/media#media.
- Source code: http://www2.scut.edu.cn/_upload/article/files/74/7c/392c 47d045e3948f22ded94344af/397341a2-6f56-42f3-b2c4.bd13f0fca3d8.zip.

4.2 A powerful approach to configure software products— "Multiobjective Evolutionary Algorithm + Estimation of Distribution"

4.2.1 Survey of advancements in research

Illustrated in Fig. 4.6, numerous renowned domestic and international enterprises, including Huawei, Boeing, Siemens, and Toshiba, have embraced Software Product Line (SPL) methodologies in their product development processes. This approach not only enhances the ability to cater to varied end-user requirements and accomplish product customization but also contributes to cost-effective development, decreased maintenance efforts, and expedited time-to-market cycles. Prominent examples of software products developed through product line technology encompass the Linux operating system, Eclipse IDE, Drupal web development system, Amazon, etc.

Indeed, SPL technology employs reusable modular software components to streamline the creation of software product sets distinguished by feature models (FMs). A feature, in this context, denotes a specific system function, while a

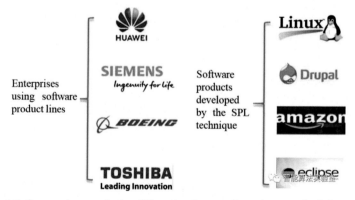

Figure 4.6 Companies employing SPL technology and products crafted through SPL technology. *SPL*, Software Product Line.

product encompasses a compilation of features. The FM articulates the constraint relationships among features, thus delineating all possible combinations that constitute an effective software product. Recently, the SPL configuration quandary has emerged as a significant focal point in SPL research, serving as a notable representative of Search-Based Software Engineering (SBSE). The SPL configuration process relies on the FM and the constraint relationships between features. It entails selecting a subset of features to optimize one or more objectives for software engineers or end users. These objectives may involve minimizing overall costs, maximizing the enhancement of chosen modules, or minimizing software defects. From a mathematical perspective, the SPL problem can be framed as an extensive, binary multiobjective optimization problem with constraints. In practical SPL scenarios, the number of features is typically vast, ranging from thousands to tens of thousands, with numerous intricate dependencies or exclusions between them, numbering from hundreds to thousands. Given this expansive and highly constrained decision space, manual configuration of software products becomes unfeasible.

To address the mentioned challenges, this section introduces an evolutionary algorithm known as MOEA/D-estimation of distribution (EoD) [29] for the configuration of SPLs. Within the decision space, a probabilistic model is implemented to estimate solution distributions, facilitating the generation of new individuals through sampling. Repair operators and constraint-handling mechanisms are devised based on domain knowledge, incorporating satisfiability solvers. In the experimental phase, simulation experiments are conducted using practical SPLs widely employed in both academic and industrial settings. The findings demonstrate the superiority of the proposed method over other prevalent approaches. These research outcomes exemplify a successful application of decomposition-based MOEAs, offering a practical solution for the automated configuration of SPLs.

4.2.2 Fundamental scientific concepts

4.2.2.1 Problem statements

In the field of SBSE, a thoroughly examined challenge pertains to the identification of the most suitable product within an SPL. To elucidate this matter distinctly, the subsequent essential concepts are initially presented.

Within an SPL, a collection of reusable modular software components is utilized to methodically construct a range of software products. Despite sharing common functionalities, these products exhibit variations in specific aspects of system functionality [30]. Typically, an SPL is depicted by an FM [31], characterized by a tree-like structure where each node signifies a feature—an abstraction of a functionality or a product characteristic [32,33]. For illustration in Fig. 4.7, the FM for an SPL related to mobile phones, consisting of 10 features, is presented.

Within an FM, every feature (excluding the root) is linked to a single parent feature but may encompass a set of child features. The FM explicitly defines both the *parent-child relations* (PCRs) and *crosstree constraints* (CTCs) among features [34]. When considering a parent feature X with its child features $\{x_1, x_2, \ldots, x_n\}$, there are four distinct types of PCRs, each expressible through a propositional formula.

- x_i is *mandatory* child feature: $x_i \leftrightarrow X$.
- x_i is *optional* child feature: $x_i \rightarrow X$.
- $\{x_1, \ldots, x_n\}$ is a group of *or* child features: $X \leftrightarrow x_1 \vee \ldots \vee x_n$.
- $\{x_1, \ldots, x_n\}$ is a group of *x or* child features: $(X \leftrightarrow x_1 \vee \ldots \vee x_n) \wedge \{\wedge_{1 \leq i < j \leq n}(\neg(x_i \wedge x_j))\}$.

Given two features x_1 and x_2, there exist the following two:

- x_1 *requires* x_2: $x_1 \rightarrow x_2$.
- x_1 *excludes* x_2: $\neg(x_1 \wedge x_2)$.

A valid product must simultaneously satisfy all the constraints derived from both PCRs and CTCs. Since any violated constraint will result in software products that are not configurable, we call them *"hard constraints"* (HCs). For example, HCs for the FM in Fig. 4.7 can be expressed in the following propositional:

$$HC = x_1 \wedge \{x_1 \leftrightarrow x_2\} \wedge \{x_1 \leftrightarrow x_4\} \wedge \{x_3 \rightarrow x_1\}$$
$$\wedge \{x_5 \rightarrow x_1\} \wedge \{x_4 \leftrightarrow xor\{x_6, x_7, x_8\}\} \qquad (4.8)$$
$$\wedge \{x_5 \leftrightarrow x_9 \vee x_{10}\} \wedge \{x_9 \rightarrow x_8\} \wedge \{\neg\{x_3 \wedge x_6\}\}$$

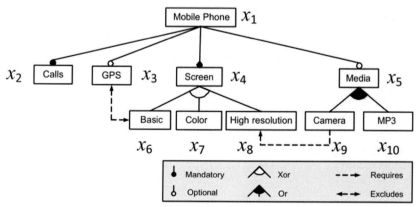

Figure 4.7 Feature model for a simplified mobile phone software product line [35].

The propositional Eq. (4.8) can be transformed into the following equivalent formula in conjunctive normal form (CNF).

$$
\begin{aligned}
HC = {} & x_1 \wedge (\neg x_1 \vee x_2) \wedge (x_1 \vee \neg x_2) \wedge (\neg x_1 \vee x_4) \\
& \wedge (x_1 \vee \neg x_4) \wedge (x_1 \vee \neg x_3) \wedge (x_1 \vee \neg x_5) \wedge (x_4 \vee \neg x_6) \\
& \wedge (x_4 \vee \neg x_7) \wedge (x_4 \vee \neg x_8) \wedge (\neg x_4 \vee x_6 \vee x_7 \vee x_8) \\
& \wedge (\neg x_6 \vee \neg x_7) \wedge (\neg x_6 \vee \neg x_8) \wedge (\neg x_7 \vee x_8) \\
& \wedge (x_5 \vee \neg x_9) \wedge (x_5 \vee \neg x_{10}) \wedge (\neg x_5 \vee x_9 \vee x_{10}) \\
& \wedge (x_8 \vee \neg x_9) \wedge (\neg x_3 \vee \neg x_6)
\end{aligned}
\tag{4.9}
$$

In Eq. (4.9), there are 19 HCs in total. Notice that each disjunctive formula in Eq. (4.9) is also called a clause.

4.2.2.1.1 The OSPS problems with soft constraints

To address the optimal software product selection (OSPS) problem, which has been extensively studied in the literature [36−39], we need to explore solutions in the decision space $\{0, 1\}^n$ by either selecting or deselecting each feature x_i ($i = 1, \ldots, n$). These solutions should meet two requirements: (1) satisfing all the HCs specified by the FM, e.g., those in Eq. (4.9), and (2) keeping a trade-off among multiple (e.g., four or more) optimization objectives, such as the total cost, the number of deselected features, and the known defects [36−39].

Previous studies, however, only considered HCs when selecting software products from an SPL. In practice, we may also need to take into account "*soft constraints*" (SCs), which are related to the demands of users. To construct multi-objective OSPS problems with SCs, we add three attributes to the ith feature, i.e., $cost_i$, $used\_before_i$, and $defects_i$, as suggested by Sayyad et al. [36,40], Henard et al. [37], and Hierons et al. [38]. The values of the attributes are randomly generated in uniform distributions. Specifically, the values of $cost_i$ are uniformly distributed between 5.0 and 15.0, while those of $defects_i$ are random integers between 0 and 10. The $used\_before_i$ takes random boolean values (represented as 0 and 1). In addition, there is a dependency between $used\_before_i$ and $defects_i$: if (not $used\_before_i$), $defects_i = 0$. It is important to note that the ranges of the attributes are selected based on the practice in the prior works [36−38,40].

Based on the above attributes, we can construct the following two-objective OSPS problem.

$$
\text{Minimize} f(x) = (f_1(x), f_2(x))
\tag{4.10}
$$

$$
\text{subject to HCs in CNF (see Section 4.2.2)}
\tag{4.11}
$$

$$\sum_{j=1}^{n} \text{cost}_j \cdot x_j \leq \sigma \sum_{j=1}^{n} \text{cost}_j \tag{4.12}$$

$$x = (x_1, \ldots, x_n)^T \in \{0, 1\}^n \tag{4.13}$$

where

$$f_1(x) = n - \sum_{j=1}^{n} x_j \tag{4.14}$$

$$f_2(x) = \sum_{j=1}^{n} x_j \cdot \left(1 - used\_before_j\right) \tag{4.15}$$

The constraint (4.11) specifies all the HCs in CNF similar to those in (4.9). The constraints in (4.12) mean that the cost of the current software product should not exceed $\sigma \sum_{j=1}^{n} \text{cost}_j$, indicating that software engineers or end users may impose budget restrictions. The parameter σ in (4.12) is set to be 0.2 in this paper.[1] According to constraint (4.13), x is an n-dimensional binary vector.

The first optimization objective $f_1(x)$, as shown in (4.14), denotes the number of unselected features. In practice, the products to be configured are expected to provide as many functionalities as possible. Therefore, we seek to maximize the number of selected features, i.e., $\sum_{j=1}^{n} x_j$. Conversely, the number of unselected features should be minimized.

The second optimization objective $f_2(x)$, as given in (4.15), represents the number of features that have not been used before. As discussed in [38,39], features that have not been used previously are more likely to be faulty. Therefore, the number of features that were not used before should be minimized. According to (4.15), we only consider features that are currently selected. If a feature is not selected, i.e., $x_j = 0$, $\left(1 - used\_before_j\right)$ contributes 0 to $f_2(x)$. Instead, if x_j is selected and was not used before, i.e., $x_j = 1$ and $\left(used\_before_j = 0\right)$, $x_j \left(1 - used\_before_j\right)$ contributes 1 to $f_2(x)$.

By adding the third optimization objective:

$$f_3(x) = \sum_{j=1}^{n} defects_j \cdot x_j \tag{4.16}$$

---

[1] According to our empirical experiments, there are both feasible and infeasible solutions by setting σ to 0.2. Therefore, the constraint (4.12) can be used to test an algorithm's ability to distinguish between feasible and infeasible solutions.

we can obtain the following three-objective OSPS:

$$Minimize\, f(x) = (f_1(x), f_2(x), f_3(x)) \tag{4.17}$$

s.t. the same constraints as in (4.11)–(4.13)

As seen, the three-objective OSPS problem has the same constraints as the two-objective problem. The added objective, given by (4.16), is the number of total defects in all the selected features. Definitely, this number should be minimized.

Finally, the four-objective OSPS problem is as follows:

$$Minimize\, f(x) = (f_1(x), f_2(x), f_3(x), f_4(x)) \tag{4.18}$$

subject to the same constraints as in Eqs. (4.11) and (4.12)

$$\sum_{j=1}^{n} defects_j \cdot x_j \leq \delta \sum_{j=1}^{n} defects_j \tag{4.19}$$

the same constraints as in Eq. (4.13)

This problem is constructed by adding the fourth optimization objective as in Eq. (4.20) and a new constraint Eq. (4.19). The fourth optimization objective denotes the total cost, which is to be minimized. The added constraint Eq. (4.19) imposes restrictions on the defects allowed. In Eq. (4.19), δ is set to be 0.1 in this work.[2]

$$f_4(x) = \sum_{j=1}^{n} cost_j \cdot x_j \tag{4.20}$$

4.2.2.2 MOEAs based on estimation of distribution

In this section, we first provide the details of the EoD update operator. Secondly, we integrate this operator into two popular decomposition-based MOEAs, i.e., MOEA/D [15] and NSGA-III [2]. Thirdly, we depict a new repair operator for the OSPS problem. Finally, we show how the soft constraints are handled in both algorithms.

The EoD update operator in most decomposition-based MOEAs, an MOP is decomposed into a number of scalar subproblems by employing N weight vectors. For each subproblem, we can build a probabilistic model using historical information, and then this model is applied to update solutions. More specifically, for the ith ($i = 1, 2, \ldots, N$)

---

[2] Similar to the parameter σ in (4–12), δ is also set based on our empirical experiments.

subproblem, we maintain a probability vector, denoted as $p_i = (p_{i1}, p_{i2}, \ldots p_{in})$, where the kth component represents the probability of being "1" (or *true*) in the kth position. In other words, $p_{ik} = P(x_{ik} = 1)$, where $P(\cdot)$ is the probability of an event. Since the exact value of p_{ik} is unknown, we use the following formula to estimate:

$$p_{i_k} = \alpha \cdot 0.5 + (1 - \alpha) \cdot \frac{T_{i_k}}{S_i} \qquad (4.21)$$

where T_{i_k} is the accumulated number of 1 in the kth gene for the ith subproblem, and S_i denotes the total number of solutions visited till now in the ith subproblem. Clearly, $\frac{T_{i_k}}{S_i} \le 1$ as $T_{i_k} \le S_i$. If we use only the ratio $\frac{T_{i_k}}{S_i}$ to estimate p_{i_k}, this would be problematic if S_i is not large enough, particularly in the early phase of the evolutionary process where a limited number of solutions have been visited.

To handle the above issues, we introduce a learning factor $\alpha \in [0, 1]$ so that two items, i.e., 0.5 and $\frac{T_{i_k}}{S_i}$, can be weighted. In the early phase, the decision variables take either 0 or 1 with a probability close to 0.5. As evolution proceeds, the term $\frac{T_{i_k}}{S_i}$ may become more and more accurate to approximate p_{i_k}. Intuitively, α is expected to be relatively large in the initial stage and gradually decrease later. Therefore, we can set α according to Eq. (4.22):

$$\alpha = 1 - \frac{FEs}{max\_FEs}, \qquad (4.22)$$

where *FEs* is the number of the current function evaluations, and *max_F Es* denotes the maximal *FEs* allowed. Similarly, if use the maximal runtime (i.e., *max_RT*) as termination condition, α is updated by

$$\alpha = 1 - \frac{RT}{max\_RT}, \qquad (4.23)$$

where *RT* is the current runtime.

According to Eqs. (4.22) and (4.23), α is linearly decreased from 1.0 to 0.0 during the whole evolutionary process. In the early phase, the second term "$\frac{T_{i_k}}{S_i}$" in Eq. (4.21) is just a coarse estimation to the distribution. Therefore, it is reasonable to give a smaller weight to it and a larger one to the first term "0.5." In other words, the p_{i_k} is close to 0.5 in the early phase, doing a random assignment between 1 and 0. However, the p_{i_k} is primarily determined by the second term because the weight 1 − α grows as α decreases in the later phase.

The EoD update operator is given in Algorithm 4.4. As shown in Line 1, the probability vector for the ith subproblem is constructed by using Eq. (4.21) to calculate each component of P_i. Similar to mutation operators, the EoD update operator introduces an update probability u_p, which determines the ratio of decision variables to be updated by EoD. For each dimension k, as shown in Line 4 of Algorithm 4.4, the EoD will be applied if a random number r_1 is smaller than u_p. To update x_k by EoD, the procedure must check if $r_2 < p_{i_k}$ is true. If so, x_k is set to be 1. Otherwise, it is set to be 0 (see Lines 6−10 in Algorithm 4.4).

Finally, it is worth mentioning that the EoD update operator can be implemented easily. From the perspective of programming, we just need to maintain a variable for S_i and an array for $T_i = \left(T_{i_1}, T_{i_2}, \ldots, T_{i_n} \right)$. In addition, once a new solution appears in the ith subproblem, variables and arrays can be automatically updated. In the next section, we will demonstrate how to integrate EoD update operators into decomposition-based MOEAs with minimal effort.

4.2.2.3 Integrating EoD into decomposition-based MOEAs

In this section, we integrate EoD into two popular decomposition-based MOEAs (i.e., MOEA/D and NSGA-III), leading to two new algorithms, namely MOEA/D-EoD and NSGA-III-EoD.

ALGORITHM 4.4 EoD_update_operator (x, i, u_p).

Input: x (the solution to be updated), i (the ith subproblem), u_p (the update probability)

Output: x (the updated solution)

1: Construct p_i according to (4.21)
2: **for** $k \leftarrow 1$ to n **do**
3:      $r1 \leftarrow rand(0, 1)$
4:      **if** $r1 < u_p$ **then**
5:          $r2 \leftarrow rand(0, 1)$
6:          **if** $r2 < p_{i_k}$ **then**
7:              $x_k \leftarrow 1$
8:          **else**
9:              $x_k \leftarrow 0$
10:          **end if**
11:      **end if**
12: **end for**
13: **return** x

The framework of the introduced MOEA/D-EoD is given in Algorithm 4.5, where the codes related to EoD are underlined. In MOEA/D-EoD, there are two common parameters N and T, which denote the population size and the neighborhood size, respectively. As in MOEA/D, a set of weight vectors $\{w_1, \ldots, w_N\}$ is needed in the new algorithm. Before the algorithm begins, the weight vectors are generated by using systematic approaches [2,25,26]. Then, the neighborhoods can be identified by calculating T closest weight vectors to each weight vector (see Lines 1–2 in Algorithm 4.5). Next, the population P is randomly initialized. If infeasible solutions appear, they are repaired according to

---

ALGORITHM 4.5 MOEA/D-EoD for binary optimization.

Input: N (population size), T (neighborhood size)
Output: The final population P
1: Initialize N weight vectors $w_1, \ldots, w_N$.
2: For each $i = 1, 2, \ldots, N$, set $B(i) = \{i_1, \ldots, i_T\}$, where $w_{i1}, \ldots, w_{iT}$ are T weight vectors closest (regarding the Euclidean distance) to w_i.
3: Generate an initial population $P = \{x_1, \ldots, x_N\}$ and repair infeasible solutions according to problem-specific methods.
4: Work out $f(x_1), \ldots, f(x_N)$.
5: **while** *the termination criterion is not fulfilled* **do**
6:     Determine z^{min} and z^{max}. For each $f(x)$, $x \in P$, normalize it to $\tilde{f}(x)$ by (4.24).
7:     Initialize the reference point z^* used in scalarizing functions.
8:     **for** $i = 1, \ldots, N$ **do**
9:         Randomly select two indexes k and l from $B(i)$, and apply crossover operator to xk and xl to generate two new solutions y_1 and y_2.
10:         Set y to y_1 or y_2, both with a probability 0.5. Similarly, set h to either k or l.
11:         _EoD_update_operator(y, h, u_p) // Algorithm 4.4_
12:         Repair y using problem-specific methods.
13:         Work out $f(y)$ and normalize it to $\tilde{f}(y)$ using the same z^{min} and z^{max} as in line 6.
14:         Update z^* based on $\tilde{f}(y)$.
15:         Update subproblems: For each $j \in B(i)$, if $g^*(y|w_j, z^*) < g^*(x_j|w_j, z^*)$, set $x_j = y$, and update S_j and T_j in EoD. // The g^* is a scalarizing function.
16:     **end for**
17: **end while**
18: **return** P

problem–specific methods. For the OSPS problem, the repair operator will be given in Section 4.2.2. The following are some important steps involved in MOEA/D-EoD.

4.2.2.3.1 Normalization and estimation of ideal/nadir points

It is recommended to normalize the objective vector (Line 6 in Algorithm 4.5). For each $f(x) = (f_1(x), ..., f_m(x))^T$, $x \in P$, we normalize it to $\tilde{f}(x) = (\tilde{f}_1(x), ..., \tilde{f}_m(x))^T$ by using $z^{min} = (z_1^{min}, ..., z_m^{min})^T$ and $z^{max} (z_1^{max}, ..., z_m^{max})^T$ according to Eq. (4.24):

$$\tilde{f}_i(x) = \frac{f_i(x) - z_i^{min}}{z_i^{max} - z_i^{min}} \tag{4.24}$$

where z_i^{min} and z_i^{max} are the minimum and maximum value found so far for the ith objective. According to Eq. (4.24), $\tilde{f}_i(x) \in [0, 1]$.

Lower and upper bounds for the objective functions are defined by the components of the ideal point z^{ideal} and the nadir point z^{nadir}, which represent the best and the worst values that each objective function can reach in the PF [16]. Since the true ideal point and nadir point are difficult to obtain, they are usually estimated during the optimization. After normalization, the ideal point can be estimated by $z^{ideal} = (-0.1, ..., -0.1)^T$, which aligns with the suggestions in [15] and [41]. According to [16], the nadir point can be estimated to be z^{max} or any objective vector dominated by it. To be consistent with the scale factor (i.e., 0.1) in the estimation of z^{ideal}, the nadir point in this section is estimated to be $z^{nadir} = (1.1, ..., 1.1)^T$.

In Line 7 of Algorithm 4.5, according to the demands of users, the reference point z^* should be initialized. If the ideal point is specified, z^* is set to be z^{ideal}. Otherwise, z^* is set to be z^{nadir}.

4.2.2.3.2 Reproduction operators

To generate new solutions, we randomly select two indexes k and l from $B(i)$ and apply a crossover operator to the parents x_k and x_l (Line 10 in Algorithm 4.5). For binary optimization, a number of crossover operators can be applied. Among them, the single-point crossover [42,43] is probably one of the simplest and most widely used operators. The single-point crossover exchanges the bits of the first parent, from the beginning to the crossover point, with those of the second one. Each time, this will generate two subsolutions y_1 and y_2. According to Line 10 in Algorithm 4.5, y is set to be either y_1 or y_2, with a same probability. That is to say, we randomly select a subsolution between y_1 and y_2, and assign it to y for further improvement through EoD. Since the genes of y come from either x_k and x_l, we use the probability vector of the kth or the l-th subproblem to update y. For this end, h is set to be either k or l, and is used as the second input parameter in the *EoD_update_operator* (Line 11).

4.2.2.3.3 Updating subproblems

Before updating the subproblems, we need to update z^* according to the type of reference point used (Line 14 in Algorithm 4.5). If the ideal point is used, z^* is updated as follows: for each $j = 1, \ldots, m$, if $\tilde{f}_j(y) < z_j^*$, set $z_j^* = \tilde{f}_j(y)$. If the nadir point is used, the update is as follows: For each $j = 1, \ldots, m$, if $\tilde{f}_j(y) < z_i^*$, set $z_j^* = \tilde{f}_j(y)$.

For each index $j \in B(i)$, both y and x_i are evaluated by a scalarizing function (i.e., g^*) with respect to w_j and the latest z^*. As shown in Line 15 of Algorithm 4.5, if $g * \left(y \middle| w_j, z^*\right) < g^*\left(x_j \middle| w_j, z^*\right)$, replace x_i with y. We should update S_j and T_j accordingly in the EoD operator because the solution of the jth subproblem is replaced by y. S_j is increased by 1. T_{jk} is increased by 1 if $y_k = 1$ as well.

In MOEA/D-EoD, the scalarizing function g^* can be the weighted sum function (g^{WS}) [15], the weighted Tchebycheff function $(g^{TCHE1}$ [15] and g^{TCHE2} [44,45]), and the penalty-based boundary intersection function (g^{PBI}) [15,41].

Finally, we provide the following comments on the introduction of MOEA/D-EoD.

- Normalization program is integrated into the framework. Therefore, the algorithm can handle problems with objective functions in different scales.
- The scalarizing functions g^{TCHE1}, g^{TCHE2}, and g^{PBI} are slightly modified such that the reference point can be either the ideal point z^{ideal} or the nadir point z^{nadir}.
- Integrating EoD into MOEA/D is easy. As shown in Algorithm 4.5, the framework of MOEA/D-EoD differs from the original MOEA/D in only two places (Lines 11 and 15).

The framework of NSGA-III-EoD is given in Algorithm 4.6. As shown in the algorithm, EoD further improves the new solution generated by the crossover operator (Lines 8 and 9). According to Line 14, the mixed population $S = P \cup Q$ undergoes the same operations as in the original NSGA-III. After this step, N promising solutions are selected from S to form the population for the next generation. For more details on the nondominated sorting, the normalization, the association, and the niche-preservation operations, refer to the original study [2].

ALGORITHM 4.6 NSGA-III-EoD for binary optimization.

Input: N (population size)

Output: The final population P

1: Initialize N reference points $z_1, ..., z_N$.

2: Generate an initial population $P = \{x_1, ..., x_N\}$ and repair infeasible solutions according to problem-specific methods.

3: **while** the termination criterion is not fulfilled **do**

4:     $Q = \varnothing$

5:     **for** $i = 1, 2, ..., N/2$ **do**

6:         Select two random indexes k, l $\{1, 2, ..., N\} \wedge k \neq l$.

7:         Apply crossover operator to x_k and x_l to generate two new solutions y_1 and y_2.

8:         EoD_update_operator(y_1, k, u_p) // Algorithm 4.4

9:         EoD_update_operator(y_2, l, u_p) // Algorithm 4.4

10:        Repair y_1 and y_2 using problem-specific methods.

11:        Add y_1 and y_2 into Q.

12:    **end for**

13:    $S = P \cup Q$

14:    Select N promising solutions from S to construct the next population P by using the nondominated sorting, the normalization, the association, and the niche-preservation operations as in the original NSGA-III.

15:    For each $j \in \{1, 2, ..., N\}$, update S_j and T_j in EoD.

16: **end while**

17: **return** P

We concern how to update S_j and T_j in the EoD operator (Line 15 of Algorithm 4.6). In NSGA-III-EoD, the objective space is divided into N subspaces by N reference lines (determined by reference points). In the selection process shown in Line 14, each solution for the next population has already been associated with a reference line based on the vertical distance and the niche count. Let the niche count for the jth reference line be ρ_j, then S_j is updated as: $S_j \leftarrow S_j + \rho_j$. The solutions associated with the jth reference line are used to update T_j. For each relevant solution, the T_{jk} is increased by 1 if the kth gene of this solution is 1.

Fig. 4.8 shows an example of updating S_j and T_j in NSGA-III-EoD. In this figure, the first and the third reference lines have only one associated solution, therefore $S_1 = S_1 + 1$ and $S_3 = S_3 + 1$. Similarly, because two solutions are associated with both the second and the forth reference lines, $S_2 = S_2 + 2$ and $S_4 = S_4 + 2$. Since there are no solutions being associated with z_5, we do not need to update S_5 and T_5.

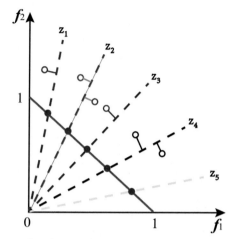

Figure 4.8 Illustration of updating S_j and T_j in NSGA-III-EoD.

Finally, it should be mentioned that EoD can be easily integrated into decomposition-based MOEAs, where an MOP is decomposed into multiple subproblems (such as MOEA/D), or the objective space is divided into multiple subspaces (such as NSGA-III). In both cases, the historical information can be easily recorded for each subproblem or each subspace, i.e., S_j and T_j, $j = 1, 2, \ldots, N$. With this in mind, the integration of EoD would be easy in decomposition-based algorithms. In this section, we provide two paradigms, namely MOEA/D-EoD and NSGA-III-EoD. In practice, EoD can also be integrated into other decomposition-based MOEAs.

4.2.2.4 A new repair operator for the OSPS problem

As shown in Algorithms 4.5 and 4.6, infeasible solutions should be fixed in both the initialization and reproduction phases. In general, the repair operator is designed based on the importance of the problems at hand. For the OSPS problem, we propose a new SAT solver-based operator to fix solutions violating HCs.

According to Section 4.2.2, HCs can be represented by CNF. In fact, finding a solution that satisfies all the HCs is essentially to solve an SAT problem. Therefore, SAT solvers can be applied naturally [39]. In the literature, there are two types of high-performance SAT solvers: conflict-driven clause learning (CDCL) algorithms [46−49] and stochastic local search (SLS) algorithms [50−53]. Inspired by the work in [39], we suggest using two types of SAT solvers simultaneously to repair infeasible solutions. The selected CDCL and SLS-type solvers are SAT4J [46] and probSAT [50], respectively. In Algorithm 4.7,

---

ALGORITHM 4.7 Repair operator for the OSPS problem.
Input: y // An infeasible solution violating HCs
Output: y // The solution after repairing
 1: **if** $rand() < \tau$ **then**
 2:    Repair y using the probSAT solver [49]
 3: **else**
 4:    Repair y using the SAT4J solver [45]
 5: **end if**
 6: **return** y

---

the repair operator for the OSPS problem is given. An infeasible solution γ (violating HCs) is repaired by either the probSAT solver [50] with a probability τ, or the SAT4J solver [46] with a probability $1 - \tau$. Because the probSAT does not explore the entire decision space, it is computationally cheaper than SAT4J, which has an exponential worst-case time complexity. However, this solver is not guaranteed to find a feasible solution starting from an infeasible solution. Therefore, the SAT4J solver is introduced to rectify the above drawback. The parameter τ controls the computational resources distributed to the two solvers, and its effect is investigated in Section 4.2.2.

To repair an infeasible solution (Line 4 in Algorithm 4.7) using the SAT4J solver, we find out the variables that are not related to the violations of constraints. Next, we retain their values and call the solver to find a feasible solution by assigning values to the remaining variables [37]. For example, consider an FM with five features and three HCs: $FM = (x_1 \vee x_5) \wedge (x_2 \vee x_3) \wedge (x_2 \vee x_5)$. The solution $y = \{0, 0, 1, 1, 0\}$ is infeasible, because it violates two constraints $(x_1 \vee x_5)$ and $(x_2 \vee x_5)$, which involves three variables x_1, x_2, and x_5. Their assignments are taken off and $y = \{k, k, 1, 1, k\}$ is made feasible in part (k is treated as any value). Afterwards, y is given to the SAT4J solver, which will complete it and return a feasible solution. For example, it may return the following solution: $y' = \{1, 1, 1, 1, 0\}$. Therefore, y is repaired to be y'.

4.2.2.5 Handling soft constraints
Any solution that satisfies all HCs can violate the SCs defined in Eqs. (4.12) and (4.19). To handle this situation, we introduce the constraint violation (CV) value [54]. The CV value of the solution x indicates the extent of the violation of constraints. In both two- and three-objective OSPS problems, there is only one SC, as shown in Eq. (4.12).

As suggested in [54], we rewrite Eq. (4.12) into the following form to normalize the constraints:

$$h(x) = \frac{\sum_{j=1}^{n} cost_j \cdot x_j}{\sigma \sum_{j=1}^{n} cost_j} - 1 \leq 0. \tag{4.25}$$

Then $CV(x) = max\{0, h(x)\}$. A larger CV value indicates more violations of the constraint. For the four-objective OSPS problem, we can similarily calculate the CV value of the solutions. Since there are two SCs, i.e., Eqs. (4.5) and (4.12), we calculate the CV value for each constraint and use the sum of these values as the final CV value. In the presence of SCs, Line 15 in Algorithm 4.5 should be modified to Algorithm 4.8; in Line 14 of Algorithm 4.6, the Pareto domination in the nondominated sorting procedure should be substitued by the constrained domination principle [54], which emphasizes small and feasible CV value solutions.

4.2.2.6 Experimental studies

In this section, we firstly present experimental results on two benchmark problems to verify the effectiveness of EoD in improving the performance of decomposition-based algorithms. Then we conduct a series of experiments on the OSPS problem to further demonstrate the improvement in EoD performance and the superiority of the improved algorithm over the latest SATVaEA [39] for the OSPS problem considered in this section.

We first specify computational settings, and then report experimental results of two-, three-, and four-objective OSPS instances.

ALGORITHM 4.8 Update subproblems considering CV values.

Input: y and B (i)

1: **for** $j \in$ B (i) **do**

2:   **if** $CV(y) < CV(x_j)$ **then**

3:     Set $x_j = y$, and update S_j and T_j in EoD.

4:   **else if** $CV(y) = CV(x_j)$ **then**

5:     **if** $(g^*(y \,|\, w_j, z^*) < g^*(x_j \,|\, w_j, z^*))$ **then**

       Set $x_j = y$, and update S_j and T_j in EoD.

6:   **end if**

7: **end for**

4.2.2.6.1 Experimental settings

In our experiments, the settings include FMs, population size and termination condition, and parameter settings.

- **Feature models.** We use 13 FMs[3] in Table 4.4 to construct instances of the OSPS problem. These models were reverse-engineered from real-world projects, such as the Linux kernel, uClinux, and FreeBSD operating systems [55]. According to [39], the FMs can be simplified by removing *mandatory and dead* features via boolean constraint propagation (BCP) [56] (for more details refer to [39]). Table 4.4 gives out the number of features and that of HCs after applying BCP. We define that large-scale FMs are those models with not less than 3000 features. As seen, most of these models are large scale, with even 28,115 features and 227,009 HCs. According to Section 4.2.2, we can construct OSPS instances with two, three, and four objectives for each model.

- **Population size and termination condition**. The population size N is set to be 100, 105, and 120 for two-, three-, and four-objective OSPS problems, respectively. Following the same practice as in [39], we use *max_RT* as the termination condition, which is set to be 6 seconds for toybox and axTLS, 30 seconds for fiasco, uClinux, and busybox-1.18.0, and 200 seconds for all the other FMs in Table 4.4. Note that these settings are independent of the number of objectives.

Table 4.4 Feature models used in this study.

FM	Features (#)	HCs (#)
Toybox	181	477
axTLS	300	1657
freebsd–icse11	1392	54,351
Fiasco	631	3314
uClinux	606	606
busybox–1.18.0	2845	12,145
2.6.28.6–icse11	6742	227,009
uClinux–config	5227	23,951
coreboot	7566	40,736
buildroot	8150	37,294
Freetz	16,481	85,671
2.6.32–2var	27,077	189,883
2.6.33.3–2var	28,115	195,815

[3] The models are available at the LVAT repository: http://code.google.com/p/linux-variability-analysis-tools.

- **Parameter settings**. In MOEA/D-EoD and MOEA/D, the neighbor size T is set to be 10, and the penalty parameter θ is set to be 5 if the PBI scalarizing function is used. Both algorithms utilize a single-point crossover operator with a crossover probability of 1.0. In MOEA/D, the bit-flip mutation is adopted, and the mutation probability is set to be 0.01. According to the results in the previous section, the update probability u_p is set to be 0.01 in MOEA/D-EoD. For SATVaEA, the same genetic operators (including parameter settings) as in MOEA/D are used in this study. Moreover, all the three algorithms adopt the same repair operator as described in Section 4.2.2. In this operator, the parameter τ is set to be 0.9.

4.2.2.6.2 Performancee metric

To evaluate algorithms, we choose the HV [22] and IGD+ [57,58] as performance metrics, and both of them can simultaneously measure convergence and diversity. The HV metric is Pareto compliant [59], while the IGD+ metric is weakly Pareto compliant [58]. Due to the above good theoretical properties, both metrics have been widely used for performance evaluations in the evolutionary multiobjective optimization. Notice that a larger HV indicates a better approximation front. For IGD+, however, a smaller value is desired.

4.2.2.6.3 Results on two-objective OSPS instances

Table 4.5 shows the average HV and IGD+ values obtained by MOEA/D-EoD (with TCHE1) and MOEA/D (with TCHE1) in some large-scale two-objective OSPS instances. As seen, MOEA/D-EoD obtains better HV values than MOEA/D on all the models except for 2.6.28.6-icse11, freetz, and 2.6.33.3-2var regarding both TCHE1 and TCHE2. As for PBI, the performance of MOEA/D-EoD is better than that of MOEA/D on five out of seven models. Regarding the IGD+ values, as shown in Table 4.5, they are in general consistent with the HV results, indicating again an improvement of MOEA/D-EoD over MOEA/D. To intuitively compare the two algorithms, we show the final solutions obtained by both algorithms on four representative FMs. As seen from Fig. 4.9, MOEA/D-EoD and MOEA/D perform similarly on 2.6.28.6-icse11 regarding both convergence and diversity. It is clear that the solutions of MOEA/D-EoD converge much better than those of MOEA/D on freebsd-icse11. For uClinux-config, the solutions obtained by MOEA/D-EoD are slightly closer to the PF than those obtained by MOEA/D, particularly at the bottom-right corner of the front. Both in terms of the

Table 4.5 Average hypervolume (HV) obtained by MOEA/D-EoD and MOEA/D on the two-objective OSPS instances.

	FM	TCHE1		TCHE2		TCHE3	
		MOEA/D-EoD	MOEA/D	MOEA/D-EoD	MOEA/D	MOEA/D-EoD	MOEA/D
HV	2.6.28.6-icse11	7.215E−01	7.221E−01	7.192E−01	7.198E−01	7.177E−01	7.116E−01
	freebsd-icse11	8.291E−01	7.278E−01	8.047E−01	7.009E−01	8.102E−01	6.780E−01
	uClinux-config	7.204E−01	7.071E−01	7.063E−01	6.971E−01	7.015E−01	6.859E−01
	buildroot	7.207E−01	7.110E−01	7.102E−01	7.063E−01	6.987E−01	6.965E−01
	freetz	7.070E−01	7.099E−01	7.038E−01	7.065E−01	6.976E−01	7.024E−01
	coreboot	7.711E−01	5.409E−01	7.605E−01	5.523E−01	7.692E−01	6.847E−01
	2.6.33.3–2var	6.784E−01	6.989E−01	6.841E−01	6.938E−01	6.927E−01	6.945E−01
IGD +	2.6.28.6-icse11	3.120E−02	3.157E−02	3.296E−02	3.243E−02	3.265E−02	3.710E−02
	freebsd-icse11	1.165E−01	2.062E−01	1.322E−01	2.175E−01	1.230E−01	2.359E−01
	uClinux-config	6.689E−02	7.946E−02	7.559E−02	8.604E−02	7.854E−02	9.368E−02
	Buildroot	3.733E−02	4.257E−02	4.340E−02	4.529E−02	5.049E−02	5.114E−02
	Freetz	2.118E−02	1.912E−02	2.287E−02	2.099E−02	2.724E−02	2.321E−02
	Coreboot	5.884E−02	1.994E−01	6.732E−02	1.945E−01	5.840E−02	1.006E−01
	2.6.33.3–2var	2.475E−02	1.610E−02	2.226E−02	1.886E−02	1.903E−02	1.856E−02

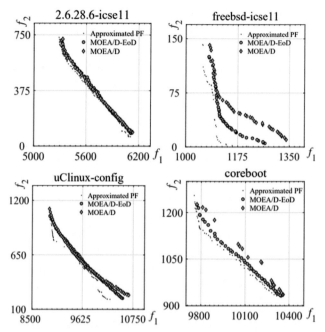

Figure 4.9 Final solutions obtained by MOEA/D-EoD and MOEA/D on four representative feature models ($m = 2$).

convergence and diversity of the final solutions, MOEA/D-EoD significantly outperforms MOEA/D on the coreboot model.

Next, we compare MOEA/D-EoD with the latest SATVaEA. Table 4.6 shows the average HV and IGD + results. As seen, MOEA/D-EoD performs much better than SATVaEA on all the models considered in terms of both HV and IGD + . As clearly shown in Fig. 4.10, the solutions of MOEA/D-EoD converge overwhelmingly better than those of SATVaEA on all the four models chosen.

The above observations lead to the following conclusion: Generally, the introduced MOEA/D-EoD outperforms MOEA/D on these large-scale two-objective OSPS instances, regardless of the scalarizing functions and reference points. In addition, MOEA/D-EoD performs significantly better than SATVaEA, particularly in terms of the convergence of the returned solutions (see Fig. 4.10).

4.2.2.6.4 Results on three-objective OSPS instances

According to Table 4.7, MOEA/D using PBI can obtain better HV results than using WS, TCHE1, and TCHE2 in the three-objective OSPS

Table 4.6 Average hypervolume (HV) and inverted generational distance (IGD) + obtained by MOEA/D-EoD (TCHE1) and SATVaEA in the two-objective OSPS instances.

FM	HV		IGD +	
	MOEA/D-EoD	SATVaEA	MOEA/D-EoD	SATVaEA
2.6.28.6–icse11	7.215E − 01	4.189E − 01	3.120E − 02	9.647E − 02
freebsd–icse11	8.291E − 01	1.797E − 01	1.165E − 01	5.100E − 01
uClinux-config	7.204E − 01	3.519E − 01	6.689E − 02	1.981E − 01
buildroot	7.207E − 01	4.273E − 01	3.733E − 02	1.012E − 01
Freetz	7.070E − 01	4.712E − 01	2.118E − 02	4.390E − 02
coreboot	7.711E − 01	4.011E − 01	5.884E − 02	1.800E − 01
2.6.32–2var	6.932E − 01	3.222E − 01	2.035E − 02	9.943E − 02
2.6.33.3–2var	6.784E − 01	3.128E − 01	2.480E − 02	8.708E − 02

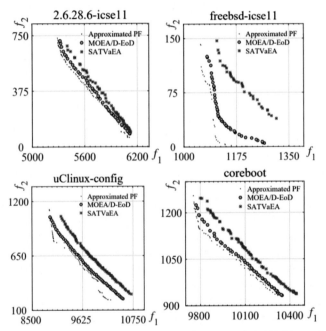

Figure 4.10 Final solutions obtained by MOEA/D-EoD and SATVaEA on four representative feature models ($m = 2$).

instances. Therefore, only PBI is used in the following comparative experiments, where MOEA/D-EoD is compared with both MOEA/D and SATVaEA. Table 4.8 gives the average HV results obtained by the three algorithms in all the three-objective OSPS instances. Compared with

Table 4.7 Average hypervolume (HV) obtained by MOEA/D (with four scalarizing functions) in the three-objective OSPS instances.

	WS	TCHE1	TCHE2	PBI
toybox	7.085E − 01	8.232E − 01	8.129E − 01	8.281E − 01
axTLS	7.120E − 01	7.685E − 01	7.607E − 01	7.818E − 01
fiasco	4.911E − 01	7.105E − 01	6.980E − 01	7.053E − 01
uClinux	7.969E − 01	8.665E − 01	8.635E − 01	8.764E − 01
busybox-1.18.0	5.151E − 01	6.354E − 01	6.156E − 01	6.533E − 01
2.6.28.6-icse11	3.127E − 01	6.376E − 01	6.073E − 01	6.529E − 01
freebsd-icse11	7.327E − 01	8.034E − 01	7.689E − 01	7.847E − 01
uClinux-config	5.267E − 01	6.836E − 01	6.618E − 01	7.112E − 01
buildroot	4.137E − 01	6.477E − 01	6.249E − 01	6.839E − 01
freetz	3.655E − 01	5.636E − 01	5.500E − 01	5.907E − 01
coreboot	4.907E − 01	6.464E − 01	6.305E − 01	6.953E − 01
2.6.32-2var	2.262E − 01	5.466E − 01	5.338E − 01	5.448E − 01
2.6.33.3-2var	2.163E − 01	5.378E − 01	5.314E − 01	5.429E − 01

Table 4.8 Average hypervolume (HV) obtained by MOEA/D-EoD (PBI), MOEA/D (PBI), and SATVaEA in the three-objective OSPS instances. The result of MOEA/D-EoD is better than (•) and equal to (‡) each of the peer algorithms.

FM	MOEA/D-EoD	MOEA/D	SATVaEA
toybox	8.259E − 01	8.281E − 01‡	6.930E − 01•
axTLS	7.840E − 01	7.818E − 01‡	6.516E − 01•
fiasco	7.026E − 01	7.053E − 01‡	6.440E − 01•
uClinux	8.797E − 01	8.764E − 01•	7.027E − 01•
busybox-1.18.0	6.656E − 01	6.533E − 01•	5.509E − 01•
2.6.28.6-icse11	6.593E − 01	6.529E − 01•	5.913E − 01•
freebsd-icse11	8.231E − 01	7.847E − 01•	6.523E − 01•
uClinux-config	7.357E − 01	7.112E − 01•	5.995E − 01•
buildroot	6.970E − 01	6.839E − 01•	6.042E − 01•
freetz	5.973E − 01	5.907E − 01‡	5.535E − 01•
coreboot	7.684E − 01	6.953E − 01•	6.612E − 01•
2.6.32-2var	5.488E − 01	5.448E − 01‡	3.923E − 01•
2.6.33.3-2var	5.415E − 01	5.429E − 01‡	3.803E − 01•

MOEA/D, MOEA/D-EoD performs equivalently on six models, i.e., toybox, axTLS, fiasco, freetz, 2.6.32-2var, and 2.6.33.3-2var, but significantly better on the remaining 7 FMs. In comparison with SATVaEA, however, MOEA/D-EoD performs significantly better on all the models considered.

According to Fig. 4.11, MOEA/D-EoD and MOEA/D can obtain similar solution sets on the small-scale toybox model. However, MOEA/D-EoD is able to find more solutions on the boundary for the two large-scale models, i.e., uClinux-config and buildroot. As for the model 2.6.33.3-2var, the solution sets found by MOEA/D-EoD and MOEA/D have no significant differences concerning the distribution of solutions. It is observed that the final solutions found by SATVaEA are mainly distributed either in the central region of the approximated PF (e.g., on toybox, uClinux-config, and buildroot) or at some extreme parts (e.g., on 2.6.33.3-2var). As given in Table 4.8, the evident underperformance of SATVaEA compared to MOEA/D-EoD concerning the HV results can mainly be attributed to the poor diversity of solutions. As seen, MOEA/D can achieve comparable performance to MOEA/D-EoD on small-scale FMs, e.g., toybox, fiasco, and on some large-scale ones, e.g., 2.6.32-2var and 2.6.33.3- 2var. In fact, these small-scale FMs are relatively easy to handle and can be well handled by both algorithms. For example, we can observe that both MOEA/D-EoD and MOEA/D can discover a high-quality solution set in the first plot of Fig. 4.11.

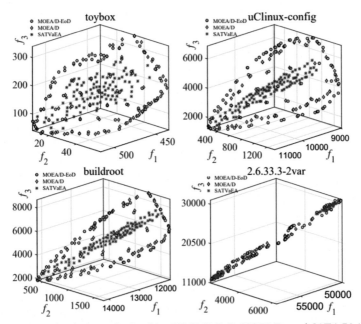

Figure 4.11 Final solutions obtained by MOEA/D-EoD, MOEA/D, and SATVaEA in four representative three-objective OSPS instances.

Large-scale 2.6.32-2var and 2.6.33.3-2var models represent the Linux kernel configuration options for the x86 architecture[4]; therefore, they have similarities in the FM trees [55]. As shown in the fourth plot of Fig. 4.11, the PF of the OSPS problem constructed based on 2.6.33.3-2var may be narrow, which could make it difficult to distinguish between solution sets found by the two algorithms.

4.2.2.6.5 Results in four-objective OSPS instances

Similarly, the HV and IGD + results in the four-objective OSPS instances are given in Table 4.9. Regarding HV, MOEA/D-EoD outperforms, matches, and underperforms MOEA/D on 7, 1, and 5 out of 13 FMs, respectively. We can observe that MOEA/D obtains comparable or even better results mainly on the first three small-scale models or on those large-scale models representing the Linux kernel configuration options for the x86 architecture (i.e., 2.6.28.6-icse11, 2.6.32-2var, and 2.6.33.3-2var). Compared with SATVaEA, according to Table 4.9, we can find that MOEA/D-EoD performs significantly better on all the models except for coreboot. Regarding IGD + , MOEA/D-EoD performs better than or equivalently to MOEA/D on all the FMs, except for 2.6.33.3-2var. Moreover, compared with SATVaEA, MOEA/D-EoD shows a significant improvement on all the models except for coreboot, on which the two algorithms yield similar performance.

4.2.2.7 Further discussions

In the previous section, we have empirically demonstrated that the EoD operator can significantly improve MOEA/D on the OSPS problems. In this section, we are going to answer the following three research questions (RQs).

- **RQ1**: How does the parameter τ affect the performance of the repair operator for the OSPS problem?
- **RQ2**: In the EoD operator, does historical information contribute more than neighboring information?
- **RQ3**: Can the EoD operator improve other decomposition-based algorithms in addition to MOEA/D?

Answers to RQ1: To answer this question, we choose MOEA/D-EoD (TCHE1 + Ideal) as an exemplary implementation, and change τ

---

[4] In fact, the 2.6.28.6-icse11 model also represents the Linux kernel configuration options [114].

Table 4.9 Average hypervolume (HV) and inverted generational distance (IGD) + for MOEA/D-EoD, MOEA/D, and SATVaEA on the four-objective OSPS problems.The result of MOEA/D-EoD is better than (•), worse than (○), and equal to (‡) each of the peer algorithms.

FM	HV			IGD +		
	MOEA/D-EoD	MOEA/D	SATVaEA	MOEA/D-EoD	MOEA/D	SATVaEA
toybox	$6.479E - 01$	$6.507E - 01○$	$5.355E - 01•$	$3.220E - 02$	$3.136E - 02‡$	$9.283E - 02•$
axTLS	$6.155E - 01$	$6.203E - 01○$	$4.938E - 01•$	$3.432E - 02$	$3.445E - 02‡$	$1.057E - 01•$
fiasco	$5.661E - 01$	$5.681E - 01○$	$5.056E - 01•$	$2.021E - 02$	$2.117E - 02‡$	$4.885E - 02•$
uClinux	$6.862E - 01$	$6.802E - 01•$	$5.401E - 01•$	$1.728E - 02$	$2.066E - 02•$	$1.334E - 01•$
busybox–1.18.0	$5.773E - 01$	$5.647E - 01•$	$4.738E - 01•$	$2.970E - 02$	$3.301E - 02•$	$9.412E - 02•$
2.6.28.6–icse11	$4.996E - 01$	$5.060E - 01○$	$4.809E - 01•$	$2.410Es - 02$	$2.283E - 02‡$	$4.607E - 02•$
freebsd–icse11	$6.170E - 01$	$5.687E - 01•$	$4.115E - 01•$	$3.513E - 02$	$5.130E - 02•$	$1.488E - 01•$
uClinux–config	$5.887E - 01$	$5.611E - 01•$	$4.920E - 01•$	$2.108E - 02$	$3.167E - 02•$	$8.394E - 02•$
buildroot	$5.437E - 01$	$5.334E - 01•$	$4.895E - 01•$	$1.814E - 02$	$2.190E - 02•$	$6.331E - 02•$
freetz	$5.136E - 01$	$5.100E - 01•$	$4.917E - 01•$	$1.370E - 02$	$1.504E - 02•$	$3.611E - 02•$
coreboot	$3.986E - 01$	$2.309E - 01•$	$4.888E - 01○$	$3.452E - 02$	$5.064E - 02•$	$2.735E - 02‡$
2.6.32–2var	$4.669E - 01$	$4.731E - 01‡$	$3.183E - 01•$	$1.501E - 02$	$1.498E - 02‡$	$1.001E - 01•$
2.6.33.3–2var	$4.653E - 01$	$4.760E - 01○$	$3.191E - 01•$	$2.107E - 02$	$1.571E - 02○$	$1.080E - 01•$

from 0.0 to 1.0 with a step size 0.1. Each of the values is tested in three two-objective OSPS instances, i.e., axTLS, fiasco, and 2.6.28.6-icse11. The three models are picked according to different termination conditions, being 6, 30, and 200 seconds, respectively. Fig. 4.12 shows the HV results over 30 runs in boxplots. According to Fig. 4.12A and B, the HV increases obviously as τ is changed from 0.0 to 0.5, and keeps steady when $\tau \in \{0.5, \ldots, 0.9\}$, but decreases when τ is switched from 0.9 to 1.0. For the 2.6.28.6-icse11 model, it is observed from Fig. 4.12C that the median of the HV values tends to rise as τ increases. Compared with $\tau = 0.9$, however, the variation range for $\tau = 1.0$ is relatively larger.

According to the above discussions, large values of τ are preferred in the repair operator, suggesting that more computational resources are

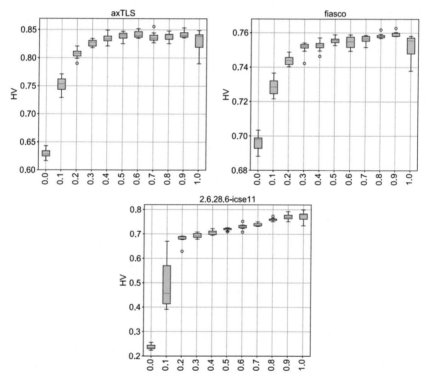

Figure 4.12 The HV values, shown in boxplots, are obtained by MOEA/D-EoD (TCHE1 + Ideal) under different values of τ. *HV*, Hypervolume.

given to probSAT than SAT4J. This is in fact consistent with the findings in [39]. Since $\tau = 0.9$ performs well in general, we recommend this parameter setting in practice.

Answers to RQ2: In the EoD operator, the probability vector is learned from historical information of each subproblem (see Eq. 4.21). In most of the previous works, however, the probability vector is constructed based on neighboring information. For example, the component of p_i in [60] is calculated:

$$ p_{ik} = \frac{\sum_{j=1}^{T} x_k^{ij} + \xi}{T + 2\xi}, \tag{4.26} $$

where x_k^{ij} denotes the kth gene of the ij subproblem (i.e., the jth neighbor of the ith subproblem), and ξ is a small value defined by $\xi = \frac{T \cdot S}{n - 2s}$. Here, s is a control parameter, which is set to be 0.4 [60]. According to Eq. (4.26), p_{ik} is determined by the neighboring information of the ith subproblem.

Table 4.10 gives the IGD + results obtained by MOEA/D-EoD (TCHE1 + Ideal), in which the probability vector is learnt from either the historical information (Eq. 4.21) or the neighboring information (Eq. 4.26). In comparison with the MOEA/D-EoD using neighboring information, the algorithm using historical information performs significantly better on all MOKP instances, all two-objective mUBQP instances, and 9 out of 13 OSPS instances. However, historical information-based MOEA/D-EoD underperforms the neighboring information-based algorithm on three small-scale OSPS instances, i.e., toybox, axTLS, and fiasco.

The above results clearly suggest that the historical information indeed contributes more than the neighboring information in the EoD operator. Due to the relatively small neighbor size, the probability vector constructed based on neighboring information is unlikely to accurately estimate the real distribution of solutions. This could be one possible reason for the ineffectiveness of neighboring information.

Answers to RQ3: As is shown in the previous section, EoD can be integrated into NSGA-III. In this section, we will answer RQ3 by comparing NSGA-III-EoD with NSGA-III on both benchmark problems and OSPS problems. The experimental results are given in Tables 4.11−4.13. As seen, NSGA-III-EoD performs better than or at least equivalently to

Table 4.10 Average inverted generational distance (IGD) + obtained by MOEA/D-EoD (TCHE1 + Ideal) using historical information and neighboring information. The result of Historical-inf is better than (•), worse than (○), and equal to (‡) Neiboring-inf.

Problem	Instance	m	Historical-inf	Neighboring-inf
MOKP mUBQP	2500	2	1.526E − 02	7.519E − 01•
	4500	4	8.701E − 02	6.364E − 01•
	6500	6	1.353E − 01	7.192E − 01•
	8500	8	1.383E − 01	8.183E − 01•
	10,500	10	1.398E − 01	1.063E + 00•
	(−0.5,1000)	2	4.315E − 02	3.461E − 01•
	(−0.2,1000)	2	6.846E − 02	5.761E − 01•
	(0.0,1000)	2	6.291E − 02	8.435E − 01•
	(0.2,1000)	2	7.879E − 02	1.239E + 00•
	(0.5,1000)	2	1.089E − 01	2.611E + 00•
	(−0.5,2000)	2	6.624E − 02	3.552E − 01•
	(−0.2,2000)	2	7.258E − 02	6.394E − 01•
	(0.0,2000)	2	1.125E − 01	9.169E − 01•
	(0.2,2000)	2	1.801E − 01	1.674E + 00•
	(0.5,2000)	2	3.370E − 01	4.307E + 00•
	(−0.5,3000)	2	7.293E − 02	3.927E − 01•
	(−0.2,3000)	2	9.749E − 02	6.794E − 01•
	(0.0,3000)	2	1.227E − 01	9.725E − 01•
	(0.2,3000)	2	1.844E − 01	1.413E + 00•
	(0.5,3000)	2	5.188E − 01	8.232E + 00•
	(−0.5,4000)	2	7.766E − 02	3.486E − 01•
	(−0.2,4000)	2	1.108E − 01	5.973E − 01•
	(0.0,4000)	2	1.233E − 01	8.021E − 01•
	(0.2,4000)	2	2.609E − 01	1.800E + 00•
	(0.5,4000)	2	5.840E − 01	5.241E + 00•
OSPS	toybox	3	4.939E − 02	4.031E − 02○
	axTLS	3	6.195E − 02	4.955E − 02○
	fiasco	3	3.670E − 02	3.251E − 02○
	uClinux	3	4.217E − 02	4.959E − 02•
	busybox−1.18.0	3	4.472E − 02	4.587E − 02•
	2.6.28.6−icse11	3	3.099E − 02	3.267E − 02•
	freebsd−icse11	3	5.418E − 02	6.254E − 02•
	uClinux−config	3	4.428E − 02	4.565E − 02•
	buildroot	3	3.695E − 02	3.680E − 02‡
	freetz	3	2.390E − 02	2.451E − 02•
	coreboot	3	4.385E − 02	4.761E − 02•
	2.6.32−2var	3	2.285E − 02	4.737E − 02•
	2.6.33.3−2var	3	2.056E − 02	5.727E − 02•

Table 4.11 Average hypervolume (HV) obtained by NSGA-III-EoD and NSGA-III on the MOKP problems. The result of NSGA-III-EoD is better than (•) and equal to (‡) NSGA-III.

	NSGA-III-EoD	NSGA-III
2500	$7.775E - 01$	$7.838E - 01‡$
4500	$1.899E - 01$	$1.595E - 01•$
6500	$4.258E - 02$	$2.472E - 02•$
8500	$1.536E - 02$	$1.047E - 02•$
10,500	$6.608E - 03$	$5.054E - 03•$

Table 4.12 Average hypervolume (HV) obtained by NSGA-III-EoD and NSGA-III on the mUBQP problems. The result of NSGA-III-EoD is better than (•) and equal to (‡) NSGA-III.

N	ρ	$m = 2$		$m = 3$	
		NSGA-III-EoD	NSGA-III	NSGA-III-EoD	NSGA-III
1000	−0.5	$6.500E - 01$	$5.950E - 01•$	$2.458E - 01$	$2.406E - 01•$
1000	−0.2	$6.301E - 01$	$5.215E - 01•$	$3.895E - 01$	$3.103E - 01•$
1000	0.0	$6.238E - 01$	$4.543E - 01•$	$4.104E - 01$	$2.601E - 01•$
1000	+0.2	$6.132E - 01$	$3.594E - 01•$	$3.873E - 01$	$1.620E - 01•$
1000	+0.5	$4.836E - 01$	$1.011E - 01•$	$1.444E - 01$	$6.500E - 06•$
2000	−0.5	$5.821E - 01$	$5.231E - 01•$	$4.114E - 01$	$4.086E - 01‡$
2000	−0.2	$5.382E - 01$	$4.268E - 01•$	$3.915E - 01$	$3.154E - 01•$
2000	0.0	$5.218E - 01$	$3.461E - 01•$	$3.332E - 01$	$1.701E - 01•$
2000	+0.2	$3.397E - 01$	$9.536E - 02•$	$1.802E - 01$	$2.082E - 02•$
2000	+0.5	$1.359E - 02$	$0.000E + 00•$	$0.000E + 00$	$0.000E + 00‡$
3000	−0.5	$5.086E - 01$	$4.872E - 01•$	$4.273E - 01$	$4.269E - 01‡$
3000	−0.2	$4.474E - 01$	$3.772E - 01•$	$3.565E - 01$	$3.012E - 01•$
3000	0.0	$3.582E - 01$	$2.690E - 01•$	$2.654E - 01$	$1.529E - 01•$
3000	+0.2	$2.702E - 01$	$1.621E - 01•$	$1.052E - 01$	$1.713E - 02•$
3000	+0.5	$0.000E + 00$	$0.000E + 00‡$	$0.000E + 00$	$0.000E + 00‡$
4000	−0.5	$5.039E - 01$	$5.054E - 01‡$	$4.526E - 01$	$4.525E - 01‡$
4000	−0.2	$4.140E - 01$	$3.961E - 01‡$	$3.073E - 01$	$2.759E - 01•$
4000	0.0	$3.752E - 01$	$3.520E - 01•$	$2.068E - 01$	$1.485E - 01•$
4000	+0.2	$5.999E - 02$	$2.780E - 02•$	$2.801E - 02$	$2.718E - 03•$
4000	+0.5	$0.000E + 00$	$0.000E + 00‡$	$0.000E + 00$	$0.000E + 00‡$

NSGA-III on all the MOKP, mUBQP, and OSPS instances. Therefore, the answer to RQ3 is clear: The EoD operator can indeed improve other decomposition-based algorithms in addition to MOEA/D. In fact, the NSGA-III is one of such algorithms.

Table 4.13 Average hypervolume (HV) obtained by NSGA-III-EoD and NSGA-III on the Software Product Line configuration problems ($m \in \{2, 3, 4\}$). The result of NSGA-III-EoD is better than (•) and equal to (‡) NSGA-III.

FM	$m = 2$		$m = 3$		$m = 4$	
	NSGA-III-EoD	NSGA-III	NSGA-III-EoD	NSGA-III	NSGA-III-EoD	NSGA-III
2.6.28.6-icse11	$7.187E - 01$	$7.214E - 01$‡	$6.147E - 01$	$6.162E - 01$‡	$4.752E - 01$	$4.757E - 01$‡
freebsd-icse11	$6.788E - 01$	$6.096E - 01$•	$7.347E - 01$	$7.388E - 01$‡	$4.338E - 01$	$4.287E - 01$‡
uClinux-config	$6.900E - 01$	$6.797E - 01$•	$6.163E - 01$	$6.151E - 01$‡	$4.683E - 01$	$4.646E - 01$‡
buildroot	$7.102E - 01$	$7.062E - 01$‡	$6.101E - 01$	$6.051E - 01$•	$4.679E - 01$	$4.671E - 01$‡
freetz	$7.100E - 01$	$7.080E - 01$‡	$5.561E - 01$	$5.560E - 01$‡	$4.767E - 01$	$4.737E - 01$‡
coreboot	$8.074E - 01$	$7.998E - 01$‡	$7.275E - 01$	$7.268E - 01$‡	$4.962E - 01$	$4.952E - 01$‡
2.6.32-2var	$6.762E - 01$	$6.876E - 01$‡	$5.417E - 01$	$5.410E - 01$‡	$3.807E - 01$	$3.888E - 01$‡
2.6.33.3-2var	$6.806E - 01$	$6.708E - 01$‡	$5.181E - 01$	$5.255E - 01$‡	$3.921E - 01$	$3.879E - 01$‡

4.2.3 Summary

This section introduces a specific application of intelligence algorithms in the field of software engineering. A decomposition-based multiobjective evolutionary algorithm is used to solve the software product configuration problem. We propose an update operator, which is integrated with two well-known decomposition-based algorithms, for enhancing decomposition-based algorithm. In addition, the multiobjective software product configuration problem with constraints is proposed. Simulation experiments are conducted on real SPLs, and the results show that the newly introduced update operator improves the performance of the decomposition-type algorithm, especially the convergence of the algorithm. Additional materials include:

- Experimental data: https://ieeexplore.ieee.org/ielx7/4235/4358751/8735924/EoD_SupplementaryMaterials.pdf?tp = &arnumber = 8735924.
- Source code: http://www2.scut.edu.cn/_upload/article/files/74/7c/392c47d045e3948f22ded94344af/6b01c219-6c4a-4984-87c8-d51920b662f8.zip.

References

[1] Xiang Y, Zhou Y, Yang X, et al. A many-objective evolutionary algorithm with Pareto-adaptive reference points. IEEE Transactions on Evolutionary Computation 2020;24(1):99−113. Available from: https://ieeexplore.ieee.org/document/8682100.

[2] Deb K, Jain H. An evolutionary many-objective optimization algorithm using reference-point based nondominated sorting approach, Part I: solving problems with box constraints. IEEE Transactions on Evolutionary Computation 2014;18 (4):577−601.

[3] Xiang Y, Zhou Y, Li M, et al. A vector angle based evolutionary algorithm for unconstrained many objective problems. IEEE Transactions on Evolutionary Computation 2017;21(1):131−52.

[4] Yuan Y, Xu H, Wang B, et al. A new dominance relation based evolutionary algorithm for many objective optimization. IEEE Transactions on Evolutionary Computation 2016;20(1):16−37.

[5] Ikeda K, Kita H, Kobayashi S. Failure of Pareto-based MOEAs: does non-dominated really mean near to optimal? In: Proceedings of the 2001 Congress on Evolutionary Computation, 2001. 2: 957−62.

[6] Li M, Yang S, Liu X. Pareto or non-Pareto: bi-criterion evolution in multiobjective optimization. IEEE Transactions on Evolutionary Computation 2016;20(5):645−65.

[7] Bhattacharjee KS, Singh HK, Ray T, et al. Decomposition based evolutionary algorithm with a dual set of reference vectors. In: 2017 IEEE Congress on Evolutionary Computation (CEC), 2017; 105−12.

[8] Liu Y, Gong D, Sun J, et al. A many-objective evolutionary algorithm using a one-by-one selection strategy. IEEE Transactions on Cybernetics 2017;47(9):2689−702.

[9] Deb K, Sundar J. Reference point based multi-objective optimization using evolutionary algorithms. In: Proceedings of the 8th Annual Conference on Genetic and Evolutionary Computation. 2006; 635−42.

[10] Zhou Y, Xiang Y, Chen Z, et al. A scalar projection and angle based evolutionary algorithm for many-objective optimization problems. IEEE Transactions on Cybernetics 2019;49(6):2073—84.

[11] Denysiuk R, Gaspar-Cunha A. Multiobjective evolutionary algorithm based on vector angle neighborhood. Swarm and Evolutionary Computation 2017;37:45—57.

[12] Cai X, Yang Z, Fan Z, et al. Decomposition-based-sorting and angle-based-selection for evolutionary multiobjective and many-objective optimization. IEEE Transactions on Cybernetics 2017;47(9):2824—37. Available from: https://ieeexplore.ieee.org/document/7516650.

[13] Cheng R, Jin Y, Olhofer M, et al. A reference vector guided evolutionary algorithm for many objective optimization. IEEE Transactions on Evolutionary Computation 2016;20(5):773—91.

[14] Liu Z, Wang Y, Huang PQ. AnD: a many-objective evolutionary algorithm with angle-based selection and shift-based density estimation. Information Sciences 2020;509:400—19.

[15] Zhang Q, Li H. MOEA/D: a multiobjective evolutionary algorithm based on decomposition. IEEE Transactions on Evolutionary Computation 2007;11(6):712—31.

[16] Saborido R, Ruiz AB, Luque M. Global WASF-GA: an evolutionary algorithm in multiobjective optimization to approximate the whole Pareto optimal front. Evolutionary Computation 2017;25(2):309—49.

[17] Sato H. Inverted PBI in MOEA/D and its impact on the search performance on multi and many objective optimization. In: Proceedings of the 2014 Annual Conference on Genetic and Evolutionary Computation, 2014; 645—52.

[18] Deb K, Thiele L, Laumanns M, et al. Scalable test problems for evolutionary multi-objective optimization.[S.l.]. Springer; 2005.

[19] Ishibuchi H, Setoguchi Y, Masuda H, et al. Performance of decomposition-based many-objective algorithms strongly depends on Pareto front shapes. IEEE Transactions on Evolutionary Computation 2017;21(2):169—90.

[20] Huband S, Hingston P, Barone L, et al. A review of multiobjective test problems and a scalable test problem toolkit. IEEE Transactions on Evolutionary Computation 2006;10(5):477—506.

[21] Coello CA, Lamont GB, Veldhuizen DAV. Evolutionary algorithms for solving multiobjective problems. Publisher Springer New York, NY Edition number: 2 https://link.springer.com/book/10.1007/978-0-387-36797-2, 2007.

[22] Zitzler E, Thiele L. Multiobjective evolutionary algorithms: a comparative case study and the strength Pareto approach. IEEE Transactions on Evolutionary Computation 1999;3(4):257—71.

[23] While L, Bradstreet L, Barone L. A fast way of calculating exact hypervolumes. IEEE Transactions on Evolutionary Computation 2012;16(1):86—95.

[24] Bader J, Zitzler E. HypE: an algorithm for fast hypervolume-based many-objective optimization. Evolutionary Computation 2011;19(1):45—76.

[25] Das I, Dennis JE. Normal-boundary intersection: a new method for generating the Pareto surface in nonlinear multicriteria optimization problems. SIAM Journal on Optimization 1998;8(3):631—57.

[26] Li K, Deb K, Zhang Q, et al. An evolutionary many-objective optimization algorithm based on dominance and decomposition. IEEE Transactions on Evolutionary Computation 2015;19(5):694—716.

[27] Wilcoxon F. Individual comparisons by ranking methods. Biometrics bulletin 1945;1(6):80—3.

[28] Derrac J, Garcia S, Molina D, et al. A practical tutorial on the use of nonparametric statistical tests as a methodology for comparing evolutionary and swarm intelligence

algorithms. Swarm and Evolutionary Computation 2011;1(1):3−18. Available from: https://www.sciencedirect.com/science/article/abs/pii/S2210650211000034.

[29] Xiang Y, Yang X, Zhou Y, et al. Enhancing decomposition-based algorithms by estimation of distribution for constrained optimal software product selection. IEEE Transactions on Evolutionary Computation 2019;1−15.

[30] Clements P, Northrop L. Software product lines: Practices and patterns. Addison-Wesley Longman Publishing Co., Inc.; 2001, p. 467.

[31] Batory D. Feature models, grammars, and propositional formulas. In: Obbink H, Pohl K, eds. Proceedings of the 9th International Conference Software Product Lines, SPLC 2005. Berlin, Heidelberg: Springer Berlin Heidelberg, 2005; 7−20.

[32] Kang K.C., Cohen S.G., Hess J.A., et al. Feature-oriented domain analysis (FODA) feasibility study. CMU/SEI-90-TR-21. Software Engineering Institute. 1990. Available from: https://insights.sei.cmu.edu/library/feature-oriented-domain-analysis-foda-feasibility-study/.

[33] Berger T, Lettner D, Rubin J, et al. What is a feature?: a qualitative study of features in industrial software product lines. In: Proceedings of the 19th International Conference on Software Product Line (SPLC), 2015; 16−25.

[34] Czarnecki K, Eisenecker U. Generative programming: methods, tools, and applications. Addison-Wesley; 2000.

[35] Benavides D, Segura S, Ruiz-Cortés A. Automated analysis of feature models 20 years later: A literature review. Information Systems 2010;35(6):615−36.

[36] Sayyad AS, Menzies T, Ammar H. On the value of user preferences in search-based software engineering: A case study in software product lines. In: 2013 35th International Conference on Software Engineering (ICSE), 2013; 492−501.

[37] Henard C, Papadakis M, Harman M, et al. Combining Multi-Objective Search and Constraint Solving for Configuring Large Software Product Lines. In: IEEE/ACM 37th IEEE International Conference on Software Engineering, 2015. 1: 517−28.

[38] Hierons RM, Li M, Liu X, et al. SIP: Optimal product selection from feature models using many objective evolutionary optimization. ACM Transactions on Software Engineering and Methodology 2016;25(2):17:1−17:39.

[39] Xiang Y, Zhou Y, Zheng Z, et al. Configuring software product lines by combining many-objective optimization and SAT solvers. ACM Transactions on Software Engineering and Methodology 2018;26(4):14:1−14:46.

[40] Sayyad AS, Ingram J, Menzies T, et al. Scalable product line configuration: A straw to break the camel's back. In: 2013 28th IEEE/ACM International Conference on Automated Software Engineering (ASE), 2013; 465−74.

[41] Ishibuchi H, Akedo N, Nojima Y. Behavior of multiobjective evolutionary algorithms on many objective knapsack problems. IEEE Transactions on Evolutionary Computation 2015;19(2):264−83.

[42] Srinivas M, Patnaik LM. Genetic algorithms: a survey. Computer 1994;27(6):17−26.

[43] Z M. Genetic algorithms + data structures = evolution programs. Springer Science & Business Media; 2013.

[44] Qi Y, Ma X, Liu F, et al. MOEA/D with adaptive weight adjustment. Evolutionary Computation 2014;22(2):231−64.

[45] Li K, Zhang Q, Kwong S, et al. Stable matching based selection in evlutionary multiobjective optimization. IEEE Transactions on Evolutionary Computation 2014;18 (6):909−23.

[46] Berre DL, Parrain A. Journal on Satisfiability, Boolean Modeling and Computation, 7 2−3, The Sat4j library, release 2.2, system description https://content.iospress.com/articles/journal-on-satisfiability-boolean-modeling-and-computation/sat190075-64.

[47] Marques-Silva JP, Sakallah, et al. GRASP: a search algorithm for propositional satisfiability. IEEE Transactions on Computers 1999;48(5):506−21.
[48] Eén N., Sörensson N. An Extensible SAT-solver. In: International conference on theory and applications of satisfiability testing. Springer Berlin Heidelberg, 2003; 502−18.
[49] Biere A. PicoSAT essentials. Journal on Satisfiability Boolean Modeling & Computation 2008;4(2−4):75−97.
[50] Balint A, Schöning U. Available from: https://link.springer.com/chapter/10.1007/ 978-3-642-31612-8_3. International Conference on Theory and Applications of Satisfiability Testing, 2012: 16−29.
[51] Selman B., Kautz H.A., Cohen B. Noise Strategies for Improving Local Search[A]. In: Proceedings of the Twelfth National Conference on Artificial Intelligence (Vol. 1). Menlo Park, CA, USA: American Association for Artificial Intelligence. 1994. 337−43. https://dl.acm.org/doi/10.5555/2891730.2891782.
[52] Lardeux F, Saubion F, Hao JK. GASAT: a genetic local search algorithm for the satisfiability problem. Evolutionary Computation 2006;14(2):223−53.
[53] Luo C, Cai S, Su K, et al. CCEHC: an efficient local search algorithm for weighted partial maximum satisfiability. Artificial Intelligence 2017;243:26−44.
[54] Jain H, Deb K. An evolutionary many-objective optimization algorithm using reference-point based nondominated sorting approach, Part II: Handling constraints and extending to an adaptive approach. IEEE Transactions on Evolutionary Computation 2014;18(4):602−22.
[55] She S, Lotufo R, Berger T, et al. Reverse engineering feature models. In: Proceedings of the 33rd International Conference on Software Engineering. 2011. Pages 461−70. https://dl.acm.org/doi/10.1145/1985793.1985856.
[56] Zabih R, Mcallester D. A Rearrangement search strategy for determining propositional satisfiability. In: AAAI'88: Proceedings of the Seventh AAAI National Conference on Artificial Intelligence. 1988; 155−60. https://dl.acm.org/doi/ abs/10.5555/2887965.2887993.
[57] Ishibuchi H, Masuda H, Tanigaki Y, et al. Difficulties in specifying reference points to calculate the inverted generational distance for many-objective optimization problems. In: IEEE Symposium on Computational Intelligence in Multi-Criteria Decision-Making (MCDM), 2014; 170−77.
[58] Ishibuchi H, Masuda H, Tanigaki Y, Nojima Y. Modified distance calculation in generational distance and inverted generational distance. Evolutionary Multi-Criterion Optimization. Cham: Springer International Publishing; 2015, p. 110−25.
[59] Zitzler E, Brockhoff D, Thiele L. The hypervolume indicator revisited: on the design of paretocompliant indicators via weighted integration. In: Obayashi S, Deb K, Poloni C, et al., editors. Evolutionary Multi-Criterion Optimization. Berlin, Heidelberg: Springer Berlin Heidelberg; 2007, p. 862−76.
[60] Wang B., Xu H., Yuan Y. Scale adaptive reproduction operator for decomposition based estimation of distribution algorithm. In: 2015 IEEE Congress on Evolutionary Computation (CEC), 2015; 2042−49.

CHAPTER 5

A new approach to intelligent algorithms for running time complexity analysis

Despite the research of evolutionary computation has increased rapidly in recent years, the research objects mainly focus on simplified algorithms, and theoretical analysis of frontier continuous evolutionary algorithms (EAs) is currently lacking. The scant research in this area is attributed to the difficulty of theoretical analysis in the comparison-based and population-based features of EAs and the complex self-adaptive strategies of frontier algorithms [1]. One of the most difficult problems is how to derive the probability density distribution functions of the random variables associated with the optimization process of EAs.

An effective way to solve the problem is to use statistical methods to model the probability density distribution functions through sampling. Currently, statistical methods, such as the Wilcoxon rank sum test, have been applied to compare the performance of EAs [2−4]. Recently, Liu et al. extended the performance profile and data profile techniques, which are used to compare the performance of EAs by analyzing means and confidence intervals [5]. In addition, experimental tools have been used in algorithm engineering to aid theoretical studies[6]. Jägersküpper and Preuss constructed four simplified cumulative step-size adaptation derived algorithms and verified whether the performance of these four derived algorithms is similar to that of these original algorithms [7]. The literature [8] further outlines a theoretical analysis of cumulative step-size adaptation derived algorithms. Although the use of statistical methods weakens the mathematical rigor, introducing statistical experiments into the theoretical analysis approach can avoid the difficulties of deriving probability density distribution functions. Existing theoretical work can be used to analyze the state-of-the-art EAs that have been successfully applied in practical optimization problems with the help of statistical methods, which can bridge the gap between theoretical foundations and practical applications.

Intelligent Algorithms
DOI: https://doi.org/10.1016/B978-0-443-21758-6.00006-1

This chapter presents a time complexity estimation method based on the average gain model for frontier continuous EAs that are not simplified and do not carry special constraints. This method introduces statistical methods into the average gain model with the help of surface fitting techniques. This chapter also includes experiments to verify the correctness and effectiveness of the proposed method. Since EAs are often referred to as evolutionary strategies in the field of continuous optimization [9], the experiments are carried out with evolutionary strategies as a case study. In addition, the similarities and differences between the proposed method and the algorithm performance comparison method are briefly discussed.

Theoretical studies of continuous EAs have lagged behind practical applications to some extent [10,11]. Most of the theoretical work simplifies the algorithms under study, making them easier to analyze. In addition, some of the findings need to be based on specific constraints, although EAs may not satisfy these constraints in practical applications. Therefore, the running time of continuous EAs that are successfully applied in practical optimization problems needs to be studied in more depth. The sampling algorithm designed in this chapter is applicable to EAs for practical applications, and the proposed method can estimate the running time of such algorithms.

5.1 Overview of research progress

The following is an example of the standard evolutionary strategy algorithm for solving the Sphere function to show the process of using this estimation method. First, samples of the gain are collected following the steps of the estimation method sampling experiment. Fig. 5.1 shows the images of the average gain obtained from the sample points of the standard evolutionary strategy algorithm with respect to the problem size and the fitness difference. Fig. 5.2 shows the effect graph obtained by taking the logarithm of the value of the average gain with a base of 10 to better characterize the data.

After fitting to the average gain data points, we obtain the fitting results as shown in Fig. 5.3. Then the upper bound of the expected first reach time can be derived, which can obtain the time complexity of the estimated evolutionary strategy, according to the average gain model Theorem 2. Figs. 5.3 and 5.4 show the images of the sample points and

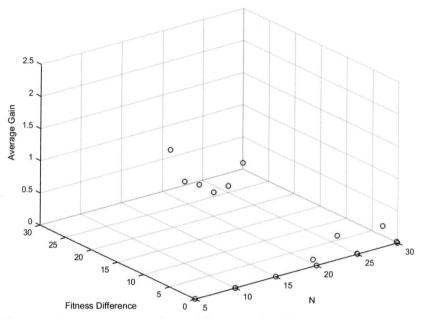

Figure 5.1 Average gain versus fitness difference and problem size.

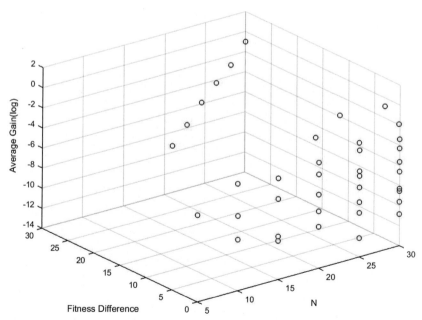

Figure 5.2 Logarithm of average gain versus fitness difference and problem size.

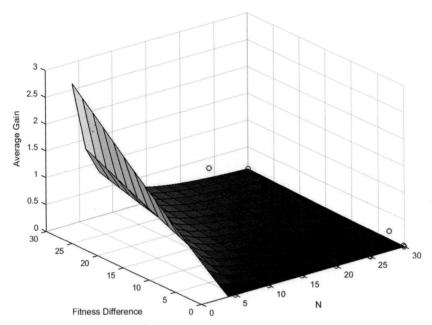

Figure 5.3 Surface fitting of average gain.

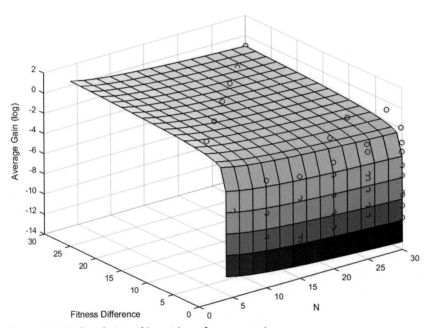

Figure 5.4 Surface fitting of logarithm of average gain.

their surface fitting results before and after taking the logarithmic values, respectively.

$$f(v, n) = \frac{1.00 \times v}{n^{1.90}}$$

$$E(T_\varepsilon | X_0) \leq 1.00 \times n^{1.90} \ln\left(\frac{X_0}{\varepsilon}\right) + 1$$

To verify the effectiveness of the proposed estimation method, this study also performs numerical experiments on the evolutionary strategy to solve the spherical function. In the experiment, the running time of the evolutionary strategy is collected to solve the spherical function in different dimensions, which is compared with the running time obtained by the estimation method. The results are presented in Fig. 5.5. The results show that the upper bound obtained from the estimation and the actual average first reach time are quite close, indicating that the estimation method possesses a high accuracy rate in analyzing the running time of the EAs.

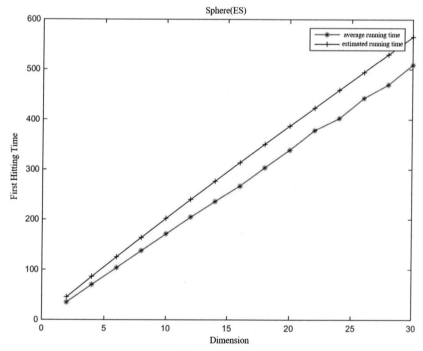

Figure 5.5 Comparison of the upper bound of estimated running time with the average running time.

In addition to the above cases, this section will apply the time complexity estimation method to estimate the running time of the simplified evolutionary strategy algorithm and the practically applied evolutionary strategy algorithm. For algorithms without available theoretical analysis results, this section will present the analysis of numerical experiments for comparison. The experimental estimation includes the typical $(1,\lambda)$ evolutionary strategy, the standard evolutionary strategy, and the self-adaptive evolutionary strategy with covariance matrix and its improved versions in the theoretical analysis cases. A total of 57 cases are included in the experiments, and the experimental results show that the estimated running time is in high agreement with the numerical experimental results.

5.1.1 Running time of $(1,\lambda)$ evolutionary strategy

In this case, Theorem 3 in the literature [12] is used to verify the running time of the estimated $(1,\lambda)$ evolutionary strategy. The algorithm implementation and parameter settings of the $(1,\lambda)$ evolutionary strategy are in accordance with the literature [12]. Since the variational operators and adaptive rules used in the literature [12] are different from those in the literature [13], the theoretical results derived in the literature [13] were not used for the validation of this experiment. The set {2, 4, 6, 8, 10, 12, 14, 16, 18, 20} is chosen as the set of the problem dimension n, and statistical experiments are conducted for each value. The nonlinear programming solver provided by MATLAB™ will be used to solve the surface fitting problem. Given that excessively large and excessively small coefficients of the estimated results would affect the estimated exponent of problem dimension n in running time, the coefficients are restricted to the range $\left[\sqrt{\frac{1}{n^*}}, \sqrt{n^*}\right]$, where n^* is the maximum value of the problem dimension in the statistical experiment. Moreover, if there is no feasible solution in this range, then the coefficients of the estimation results can be accepted for solutions in the range of $\left[\sqrt{\frac{1}{n^*}}, \sqrt{n^*}\right]$. All experiments will follow the above process for surface fitting.

The collected sample points are fitted according to the average gain collected at different values of the problem dimension n. The fitting results are as follows.

$$f(v, n) = \frac{1.56 \times v}{n^{0.64}}$$

where v denotes the adaptation value difference and n denotes the problem size. The results of estimating the upper bound on the first hitting time (FHT) for the $(1,\lambda)$ evolutionary strategy are shown below.

$$E(T_\varepsilon|X_0) \le 0.64 \times n^{0.64} \ln\left(\frac{X_0}{\varepsilon}\right) + 1$$

The conclusion in the literature [12] shows that the running time $E(T_\varepsilon|X_0) \in O\left(ln\left(\frac{X_0}{\varepsilon}\right)\right)$ for the $(1,\lambda)$ evolutionary strategy to solve the spherical function is the same as the form of the estimation, which shows the validity of the estimation method.

5.1.2 Running time of the evolutionary strategy and the covariance matrix adaptation evolutionary strategy

Since the $(1,\lambda)$ evolutionary strategy discussed above still belongs to the simplified continuous type EA, the standard evolutionary strategy (ES) and the covariance matrix adaptation evolutionary strategy (CMA-ES) [14] are chosen for this case to demonstrate the applicability of the estimation method. This case contains the time complexity estimation experiments of the standard evolutionary strategy and the covariance matrix adaptation evolutionary strategy to solve 10 benchmark test functions. The case also contains numerical experiments to compare the performance of the two algorithms to obtain the average FHT of the two algorithms in different cases. By comparing the estimation results with the numerical results, we verify whether the time complexity obtained by the estimation method can effectively reflect the performance differences of the algorithms. Eight of the 10 test functions used in the time complexity estimation experiments are from the literature [15], and the other 2 are discussed cases from the literature [14] (the Arbitrarily Oriented Hyper-Ellipsoid function and the Rosenbrock function). The algorithms are implemented according to the description in [14,15], where the problem size n is set to be 20, the population size is set to be 10, and the number of samples is set to be 100. Since [14] does not set a fixed upper limit on the number of function evaluations (FEs), the upper limit on the number of FEs is set to be 10^5.

Table 5.1 shows the estimated time complexity of the evolutionary strategy and the covariance matrix adaptation evolutionary strategy, while Table 5.2 shows the results of the corresponding numerical experiments, where n represents the problem dimension, X_0 represents the adaptation value difference of the initial solution, and ε represents

Table 5.1 Estimated time complexity and performance comparison of evolution strategy (ES) and covariance matrix adaptation evolutionary strategy (CMA-ES).

Fitness function	CMA-ES		ES		Comparison results
	Time complexity	Correctness	Time complexity	Correctness	
Cigar	$5.48 \times n^{2.38}\ln\left(\frac{X_0}{\varepsilon}\right) + 1$	Correct	$5.48 \times n^{5.24}\ln\left(\frac{X_0}{\varepsilon}\right) + 1$	Correct	CMA-ES
Different powers	$3.84 \times n^{1.68}\ln\left(\frac{X_0}{\varepsilon}\right) + 1$	Correct	$0.60 \times n^{2.62}\ln\left(\frac{X_0}{\varepsilon}\right) + 1$	Correct	CMA-ES
Ellipsoid	$30.0 \times n^{1.34}\ln\left(\frac{X_0}{\varepsilon}\right) + 1$	Correct	$30.0 \times n^{4.09}\ln\left(\frac{X_0}{\varepsilon}\right) + 1$	Correct	CMA-ES
Parabolic Ridge	$5.48 \times n^{1.16}\ln\left(\frac{X_0}{\varepsilon}\right) + 1$	Correct	$4.17 \times n^{2.00}\ln\left(\frac{X_0}{\varepsilon}\right) + 1$	Correct	CMA-ES
Schwefel	$14.7 \times n^{0.46}\ln\left(\frac{X_0}{\varepsilon}\right) + 1$	Correct	$0.10 \times n^{2.28}\ln\left(\frac{X_0}{\varepsilon}\right) + 1$	Correct	CMA-ES
Sphere	$0.18 \times n^{2.13}\ln\left(\frac{X_0}{\varepsilon}\right) + 1$	Correct	$1.00 \times n^{0.90}\ln\left(\frac{X_0}{\varepsilon}\right) + 1$	Correct	CMA-ES
Sharp ridge	$5.47 \times n^{8.62}\ln\left(\frac{X_0}{\varepsilon}\right) + 1$	Correct	$11.7 \times n^{10}\ln\left(\frac{X_0}{\varepsilon}\right) + 1$	Correct	CMA-ES
Tablet	$5.47 \times n^{3.11}\left(\frac{1}{\varepsilon^{0.02}} - \frac{1}{X_0^{0.02}}\right) + 1$	Correct	Positive infinity	Correct	CMA-ES
Arbitrarily orientated hyper-ellipsoid	$5.48 \times n^{2.09}\ln\left(\frac{X_0}{\varepsilon}\right) + 1$	Correct	$5.39 \times n^{9.87}\left(\frac{1}{\varepsilon^{0.06}} - \frac{1}{X_0^{0.06}}\right) + 1$	Correct	CMA-ES
Rosenbrock	$5.48 \times n^{2.90}\ln\left(\frac{X_0}{\varepsilon}\right) + 1$	Correct	$5.48 \times n^{5.26}\left(\frac{1}{\varepsilon^{0.39}} - \frac{1}{X_0^{0.39}}\right) + 1$	Correct	CMA-ES

Table 5.2 Average number of function evaluations (FEs) and best fitness of evolution strategy (ES) and covariance matrix adaptation evolutionary strategy (CMA-ES).

Fitness function	CMA-ES Fitness Mean	St.D.	Number of FEs Mean	St.D.	ES Fitness Mean	St.D.	Number of FEs Mean	St.D.	Comparison results	Consistency
Cigar	0.00E+00	0.00E+00	1.00E+04	2.44E+02	1.17E+01	1.68E+01	1.00E+05	0.00E+00	CMA-ES	Consistent
Different powers	0.00E+00	0.00E+00	4.57E+04	3.86E+03	1.07E−08	6.09E−09	1.00E+05	0.00E+00	CMA-ES	Consistent
Ellipsoid	0.00E+00	0.00E+00	2.62E+04	4.52E+02	5.68E+02	4.94E+02	1.00E+05	0.00E+00	CMA-ES	Consistent
Parabolic Ridge	0.00E+00	0.00E+00	1.61E+03	9.77E+02	0.00E+00	0.00E+00	8.13E+03	1.56E+04	CMA-ES	Consistent
Schwefel	0.00E+00	0.00E+00	8.21E+03	3.15E+02	0.00E+00	0.00E+00	1.54E+04	1.11E+03	CMA-ES	Consistent
Sphere	0.00E+00	0.00E+00	3.48E+03	1.26E+02	0.00E+00	0.00E+00	3.38E+03	1.64E+02	ES	Inconsistent
Sharp Ridge	0.00E+00	0.00E+00	1.23E+04	1.46E+04	6.85E−01	1.48E+00	3.24E+04	4.61E+02	CMA-ES	Consistent
Tablet	0.00E+00	0.00E+00	3.09E+04	5.96E+02	1.83E+02	1.12E+02	1.00E+05	0.00E+00	CMA-ES	Consistent
Arbitrarily Orientated Hyper-Ellipsoid	0.00E+00	0.00E+00	2.66E+04	5.24E+02	7.69E+01	4.82E+01	1.00E+05	0.00E+00	CMA-ES	Consistent
Rosenbrock	0.00E+00	0.00E+00	2.62E+04	1.27E+03	1.43E+00	1.73E+00	1.00E+05	0.00E+00	CMA-ES	Consistent

the termination threshold. The comparison results of the upper bound on the expected FHT (EFHT) in Table 5.1 are judged based on the numerical magnitude of the upper bound for the evolutionary and the covariance matrix adaptation evolutionary strategy with the same parameter settings as the numerical experiments, i.e., with $n = 20$ and $\varepsilon = 10^{-10}$. The results of the numerical experiments are determined by comparing the optimal fitness or the average evaluations of the covariance matrix adaptation evolutionary strategy. Since the global minimum of Parabolic Ridge function and Sharp Ridge function is negative infinity and the difference of adaptation values at any point in the fetching space is positive infinity, the gain cannot be calculated in these functions. Therefore, for the sake of uniformity, zero is chosen as the target adaptation value of these two functions. When the parameters in the estimation results are replaced with the parameter values set by the numerical experiments, the estimation results are said to be correct if the actual value of the upper bound of the EFHT is not smaller than the average running time observed by the numerical experiments. In addition, when at least one algorithm achieves the target accuracy within the limited running resources, the estimation result is said to be consistent with the numerical data if the algorithm with the smaller upper bound obtained from the estimation also performs well in the numerical experiment. When both algorithms fail to achieve the target accuracy within the limited running resources in the numerical experiments, the results are said to be consistent if the estimated time complexity also exceeds the upper bound on the number of iterations. Otherwise, the estimated results of the algorithms' running time and the corresponding numerical data are said to be inconsistent.

According to Table 5.2, in the case of solving nine of the benchmark test functions, the time complexity estimation results are consistent with the corresponding numerical experimental results. In the cases of solving the Schwefel function and Parabolic Ridge function, the upper bound of the EFHT of the covariance matrix adaptation evolutionary strategy is smaller than that of the evolutionary strategy, while the number of FEs is less in the former than the latter. In the cases of solving seven functions such as Cigar, Different Powers, and Ellipsoid, the optimal solution of the evolutionary strategy is worse than that of the covariance matrix adaptation evolutionary strategy, and the upper bound of time complexity of the former is larger than that of the latter.

5.1.3 Running time of the improved covariance matrix adaptation evolutionary strategy

To further demonstrate the applicability of the estimation method to the frontier evolutionary strategy algorithm, the improved covariance matrix adaptation evolutionary strategy is selected for comparison in this case [16,17], denoted as CMAES-1 and CMAES-2, respectively. This case performs the estimation experiment of time complexity with improved covariance matrix adaptive evolutionary strategy for 18 benchmark test functions and also compares the performance of the two algorithms. This section verifies whether the time complexity obtained by the estimation method can effectively reflect the differences of the algorithms. The 18 test functions used in the estimation experiments are from the literature [18]. Since the global minima of some of the benchmark test functions are not stable and may vary with the problem dimension and other factors, the benchmark test functions in this part are from the appendix of the literature [18], whose global minimum value is 0 and the function expression is fixed. The parameters of the algorithm are set as follows: the termination threshold is set to be $\varepsilon = 10^{-10}$, the problem size n is set to be 50, the population size of the parent generation is set to be 12, the population size of the children is set to be 6, the number of samples is 100, and the maximum number of iterations is set to be 5×10^4. The numerical experiments are repeated 50 times for each benchmark test function. In addition, the initial populations of CMAES-1 and CMAES-2 are always the same.

Table 5.3 shows the estimated time complexity of the two covariance matrix adaptation evolutionary strategies, and Table 5.4 presents the corresponding numerical experimental results. The comparison results in Table 5.3 are based on the EFHT upper bound for the two covariance matrix adaptive evolution strategies (ESs) estimated at $n = 50$, $\varepsilon = 10^{-10}$, while the performance comparison criteria for the numerical experiments are the same as those stated previously.

Tables 5.3 and 5.4 show that in 16 benchmark test functions the estimation results are consistent with the numerical experimental data, while the remaining two sets of estimation results do not agree with the experimental data. In addition, when the algorithm solves six benchmark test functions such as Tablet, the average gain obtained by the algorithm appears to be equal to 0 in one or more problem dimensions. This indicates that the algorithm is trapped in a local optimal solution, which

Table 5.3 Estimated time complexity and performance comparison of the two CMA-ES (covariance matrix adaptation evolution strategy).

Fitness function	CMA-ES-1 Time complexity	Correctness	CMA-ES-2 Time complexity	Correctness	Comparison results
Ackley	$5.46 \times n^{1.18}\ln\left(\frac{X_0}{\varepsilon}\right) + 1$	Correct	$4.35 \times n^{1.29}\ln\left(\frac{X_0}{\varepsilon}\right) + 1$	Correct	CMA-ES-1
Griewank	$30.0 \times n^{2.82}\ln\left(\frac{X_0}{\varepsilon}\right) + 1$	Correct	$30.0 \times n^{3.15}\ln\left(\frac{X_0}{\varepsilon}\right) + 1$	Correct	CMA-ES-1
Dixon Price	Positive infinity	Correct	Positive infinity	Correct	—
Sphere	$2.23 \times n^{1.25}\ln\left(\frac{X_0}{\varepsilon}\right) + 1$	Correct	$2.77 \times n^{0.87}\ln\left(\frac{X_0}{\varepsilon}\right) + 1$	Correct	CMA-ES-2
Schwefel	$5.21 \times n^{2.21}\ln\left(\frac{X_0}{\varepsilon}\right) + 1$	Correct	$5.48 \times n^{3.39}\left(\frac{1}{\varepsilon^{0.01}} - \frac{1}{X_0^{0.01}}\right) + 1$	Correct	CMA-ES-1
Rosenbrock	$5.48 \times n^{3.24}\ln\left(\frac{X_0}{\varepsilon}\right) + 1$	Correct	$30.0 \times n^{5.09}\ln\left(\frac{X_0}{\varepsilon}\right) + 1$	Correct	CMA-ES-1
Hyper-Ellipsoid	$0.99 \times n^{1.72}\ln\left(\frac{X_0}{\varepsilon}\right) + 1$	Correct	$0.51 \times n^{2.04}\ln\left(\frac{X_0}{\varepsilon}\right) + 1$	Correct	CMA-ES-1
Quadric	$1.03 \times n^{2.51}\ln\left(\frac{X_0}{\varepsilon}\right) + 1$	Correct	$5.48 \times n^{3.40}\left(\frac{1}{\varepsilon^{0.01}} - \frac{1}{X_0^{0.01}}\right) + 1$	Correct	CMA-ES-1
Absolute value	$0.54 \times n^{1.83}\ln\left(\frac{X_0}{\varepsilon}\right) + 1$	Correct	$1.37 \times n^{1.68}\ln\left(\frac{X_0}{\varepsilon}\right) + 1$	Correct	CMA-ES-1
Ellipsoid	$5.48 \times n^{3.79}\left(\frac{1}{\varepsilon^{0.01}} - \frac{1}{X_0^{0.01}}\right) + 1$	Correct	$5.48 \times n^{4.12}\left(\frac{1}{\varepsilon^{0.02}} - \frac{1}{X_0^{0.02}}\right) + 1$	Correct	CMA-ES-1
Quartic	$0.29 \times n^{1.75}\ln\left(\frac{X_0}{\varepsilon}\right) + 1$	Correct	$5.48 \times n^{1.10}\ln\left(\frac{X_0}{\varepsilon}\right) + 1$	Correct	CMA-ES-1
Rastrigin	Positive infinity	Correct	Positive infinity	Correct	—
Schwefel 2.22	$5.48 \times n^{1.43}\ln\left(\frac{X_0}{\varepsilon}\right) + 1$	Correct	$0.09 \times n^{2.28}\ln\left(\frac{X_0}{\varepsilon}\right) + 1$	Correct	CMA-ES-2
Step	$8.80 \times n^{0.43}\ln\left(\frac{X_0}{\varepsilon}\right) + 1$	Correct	$5.48 \times n^{1.29}\ln\left(\frac{X_0}{\varepsilon}\right) + 1$	Correct	CMA-ES-1
Schwefel 2.21	$0.18 \times n^{2.91}\ln\left(\frac{X_0}{\varepsilon}\right) + 1$	Correct	$2.58 \times n^{1.49}\ln\left(\frac{X_0}{\varepsilon}\right) + 1$	Correct	CMA-ES-2
Salomon	Positive infinity	Correct	Positive infinity	Correct	—
Schaffer6	Positive infinity	Correct	Positive infinity	Correct	—
Weierstrass	Positive infinity	Correct	Positive infinity	Correct	—

Table 5.4 Average number of function evaluations (FEs) and best fitness of the two CMA-ES (covariance matrix adaptation evolution strategy).

Fitness function	CMA-ES Fitness Mean	CMA-ES Fitness St.D.	CMA-ES Number of FEs Mean	CMA-ES Number of FEs St.D.	ES Fitness Mean	ES Fitness St.D.	ES Number of FEs Mean	ES Number of FEs St.D.	Comparison results	Consistency
Ackley	0.00E + 00	0.00E + 00	1.34E + 04	2.82E + 03	2.05E − 02	1.45E − 01	1.40E + 04	5.21E + 03	CMA-ES-1	Consistent
Griewank	1.87E-03	4.55E-03	1.76E + 04	1.43E + 04	2.07E − 03	4.06E − 03	2.02E + 04	1.60E + 04	CMA-ES-1	Consistent
Dixon price	6.67E − 01	2.39E − 16	5.00E + 04	0.00E + 00	6.67E − 01	3.44E − 16	5.00E + 04	0.00E + 00	CMA-ES-1	Consistent
Sphere	0.00E + 00	0.00E + 00	6.81E + 03	1.50E + 02	0.00E + 00	0.00E + 00	6.74E + 03	1.46E + 02	CMA-ES-2	Consistent
Schwefel	0.00E + 00	0.00E + 00	3.25E + 04	5.87E + 02	1.45E − 09	2.22E − 09	5.00E + 04	6.40E + 01	CMA-ES-1	Consistent
Rosenbrock	3.16E + 01	1.86E + 01	5.00E + 04	0.00E + 00	3.96E + 01	2.72E + 01	5.00E + 04	0.00E + 00	CMA-ES-1	Consistent
Hyper-Ellipsoid	0.00E + 00	0.00E + 00	1.18E + 04	3.51E + 02	0.00E + 00	0.00E + 00	1.19E + 04	3.80E + 02	CMA-ES-1	Consistent
Quadric	0.00E + 00	0.00E + 00	3.24E + 04	4.61E + 02	1.73E − 09	3.14E-09	5.00E + 04	1.78E + 02	CMA-ES-1	Consistent
Absolute Value	1.14E − 06	6.74E − 06	2.91E + 04	1.44E + 04	6.13E − 03	2.10E − 02	3.41E + 04	1.55E + 04	CMA-ES-1	Consistent
Ellipsoid	8.31E + 03	3.04E + 04	5.00E + 04	0.00E + 00	4.56E + 04	1.68E + 04	5.00E + 04	0.00E + 00	CMA-ES-1	Consistent
Quartic	0.00E + 00	0.00E + 00	4.31E + 03	1.49E + 02	0.00E + 00	0.00E + 00	4.18E + 03	1.57E + 02	CMA-ES-2	Inconsistent
Rastrigin	1.29E + 02	2.77E + 01	5.00E + 04	0.00E + 00	1.31E + 02	2.26E + 01	5.00E + 04	0.00E + 00	CMA-ES-1	Consistent
Schwefel 2.22	4.12E − 06	2.87E − 05	3.16E + 04	1.46E + 04	2.84E − 04	1.20E − 03	3.57E + 04	1.62E + 04	CMA-ES-1	Inconsistent
Step	0.00E + 00	0.00E + 00	4.56E + 03	2.06E + 04	1.18E + 00	1.21E + 00	3.50E + 04	2.21E + 04	CMA-ES-1	Consistent
Schwefel 2.21	8.60E + 01	3.51E + 01	5.00E + 04	0.00E + 00	1.45E − 06	4.06E − 06	5.00E + 04	0.00E + 00	CMA-ES-2	Consistent
Salomon	1.86E + 00	2.34E − 01	5.00E + 04	0.00E + 00	2.12E + 00	3.48E − 01	5.00E + 04	0.00E + 00	CMA-ES-1	Consistent
Schaffer6	2.35E + 01	3.74E − 01	5.00E + 04	0.00E + 00	2.19E + 01	6.88E − 01	5.00E + 04	0.00E + 00	CMA-ES-2	Consistent
Weierstrass	5.29E + 00	1.34E + 00	5.00E + 04	0.00E + 00	5.94E + 00	1.53E + 00	5.00E + 04	0.00E + 00	CMA-ES-1	Consistent

means that the algorithm cannot find the global optimal solution in the limited computation time. Therefore, the upper bound on the EFHT corresponding to these cases is positive infinity in Table 5.3.

The above experiments all adopt the sampling gap selection method and use the average fitting error as the judging criterion for selecting the optimal sampling gap. However, it was found that an excessively large sampling gap would lead to inaccurate estimation results. In these cases, the algorithm could achieve the target accuracy within a small number of iterations, which causes this error. For example, in CMAES-1 to solve the Hyper-Ellipsoid problem, if the sampling interval is set to be 800, the fitted result is $f(v, n) = \frac{v}{1.17 \times n^{0.60}}$. When $n = 50$, the estimated upper bound on the expected first reach time is $4.03E + 02$. However, the results of numerical experiments show that the average FHT of CMAES-1 for solving the Hyper-Ellipsoid problem with the same parameter settings is $9.83E + 02$, which is larger than $4.03E + 02$. Therefore, the estimated time complexity at a sampling interval of 800 is wrong. The reason is that when the total number of iterations is small but the sampling interval is large, the number of sample points collected will be few, and the small number of sample points may not fully reflect the variation of the average gain. Hence, the upper bound of the estimated time complexity is not used when the number of sample points is lower than 30. The experimental results show that all the estimation results are correct after using 30 as the minimum number of sample points.

5.2 Scientific principle

5.2.1 Problem description

In recent years, the running time of EAs has received extensive attention. The results about running time help deepen the understanding of EAs and evaluate the efficiency of algorithms. As a commonly used indicator to measure the running time of EAs, the FHT is the number of iterations that EAs obtain the global optimal solution for the first time [19]. Moreover, the EFHT, the average number of iterations required to find the global optimum, reflects the average time complexity of EAs [20]. Thus, the EFHT is an important concept in running time analysis.

In the last decade, running time analysis has made great advances in the field of evolutionary computation. Although there have been many studies for EAs in discrete optimization [21−23], the running time of EAs that have been successfully applied to real-world application problems in the continuous optimization, like ESs, has rarely been analyzed. Since a large number of real-world application problems are continuous, the running time of EAs in continuous optimization is of great importance. Therefore, the EFHT of EAs for continuous optimization will be discussed in this section.

In the last decade, a number of researchers have developed the running time of EAs for continuous optimization in the form of case studies, and there are many important findings. For example, Jägersküpper analyzed the running time of $(1 + 1)$ evolutionary strategy with Gaussian mutation for solving spherical functions [24], and further derived the continuous computational time of $(1 + 1)$ evolutionary strategy for solving unimodal optimization problems [25]. Beyer et al. [26] investigated the solution process of a multiple-group evolutionary strategy with σ-self-adaptation (σSA) for a specific subset of positive definite quadratic forms (PDQFs). Moreover, the scope of investigated functions was extended to the general PDQF [27]. Recently, Jiang et al. [11] derived a tighter lower bound for the $(1 + 1)$ evolutionary strategies on a Sphere function. Since the above studies mainly discuss simplified continuous EAs, there is still a lack of research on the computational time of continuous EAs for practical applications.

So far, some theoretical approaches have been proposed to investigate the running time of EAs, including fitness level method [22,28−30], drift analysis [19,31−35], and switch analysis [36,37].

In the fitness level method, the running time of the EAs is regarded as the sum of a set of waiting times, each of which represents the number of iterations to be consumed at a specific level. The method was initially proposed by Wegener [28]. The original method requires that none of these levels can be skipped, while Sudholt [22] successfully relaxed this restriction and extended the scope of studied functions to all unimodal functions. Zhou [29] incorporated tail-bound into the fitness level method. In addition, Witt [30] removed the constraint condition of the fitness level method and estimated the computation time of the random local search for solving the OneMax problem. These work analyses abstract schemes or simplified algorithms, including the randomized local optimization algorithm, $(1 + 1)$ EAs, $(\mu + 1)$ EAs, binary particle swarm algorithms, and so on. Some conclusions need to be based on assumptions about conditions that are less likely to occur in practical applications, such as the algorithm being restricted to using only bit-flip mutation operators [22]. Fitness level method has been used to analyze a number of concrete cases, but there are still few results about continuous EAs.

In addition to the fitness level method, there is another commonly used analysis method, drift analysis. Drift analysis is a general theory for the running time of EAs, which was originally proposed by He and Yao [31], and has been extended, improved, and refined by many researchers, obtaining fruitful research results. Jägersküpper [32] integrates drift analysis and Markov chain analysis. Chen et al. [33] introduced the concept of takeover time into drift analysis and discussed the time complexity of using (N + N) EA for solving the OneMax and Leadingones functions. However, unlike real EAs, the (N + N) EA does not adopt any recombination operator. To simplify the time-consuming and complicated calculation, Oliveto and Witt [35] showed how the derivation can be done in a simpler and clearer way. Apart from obtaining the boundary of average FHT, Lehre and Witt [34] used the variable drift analysis method to derive the distribution of FHT. He and Yao [19] rigorously analyzed the effect of population size on the running time of EAs using mutation and elitist selection. Theoretically drift analysis is suitable for both discrete and continuous optimization. However, since the target space is continuous or consists of a large number of continuous subspaces, research on EAs in a continuous domain is scarce [38].

Unlike the above two methods, when the switch analysis method is applied, an EA is not examined *de novo*. Instead, another EA serves as a reference to obtain better results. Yu et al. [36] proposed the method and proved that both the fitness level method and the drift analysis method can be reduced to the switch analysis method. Yu and Qian [37] further showed that the convergence-based analysis method can also be reducible to the switch analysis method and proved that the switch analysis method can yield a tighter lower bound in the case of (1 + 1) EA on the trap problem. The switch analysis method is currently discussed mainly for discrete EAs, while there are few theoretical results related to continuous EAs.

To analyze the running time of EAs for continuous optimization, Huang et al. [39] developed the average gain model, inspired by the idea of drift analysis, to estimate the average runtime of (1 + 1) EA on the Sphere function. To generalize the average gain model, Zhang et al. [12] introduced supermartingale and stopping time into the model. The concepts and theorems of the average gain model in [39] and [12] lay the theoretical basis for the proposed method.

5.2.2 Problem modeling

The following global optimization problem is considered:

$$\min f(x)$$

$$s.t. x \in M \subseteq \mathbb{R}^n$$

where the objective function $f(x)$ maps from search space M to $\mathbb{R}$ and n is the number of dimensions. Let p_{opt} denote a global optimum, and $P_t = \{p_1^{(t)}, p_2^{(t)}, \cdots, p_\lambda^{(t)}\}$ denote the offspring population at the t-th generation where λ is the population size.

Definition 5.1: Assume that p is a specific solution in the solution space and f' is a value of fitness desired to be obtained, then $d(p) = \max\{0, f(p) - f'\}$ denotes the fitness difference.

Let $\delta(P_t)$ be the smallest fitness difference obtained at the tth generation, $\delta(P_t) = \min(d(p_k^{(t)}))$, where $k = 1, 2, \cdots, \lambda$. Furthermore, let φ_t be the smallest fitness difference obtained in the previous t generations, i.e., $\varphi_t = \min(\delta(P_t))$, where $i = 1, 2, \cdots, t$.

The optimization of the objective functions by EAs can be seen as a gambling behavior, since the process of generating offspring is random. The same population is likely to produce many different offspring, whereby the corresponding fitness values may either increase or decrease. Due to its randomness, the process can be modeled by stochastic processes. Let (Ω, F, P) denote a probability space. Let $\{s_t\}_0^\infty$ be a stochastic process on (Ω, F, P), and let $F_t = \sigma(s_0, s_1, \cdots, s_t)$ be the natural filtration of F.

Definition 5.2: Given a generation t, the gain at t is given by

$$g_t = \varphi_t - \varphi_{t-1}$$

Let $H_t = \sigma\left(\varphi_0, \varphi_1, \cdots, \varphi_t\right)$, whereby the average gain at t is

$$E(g_t|H_t) = E(\varphi_t - \varphi_{t-1}|H_t)$$

The gain is equal to the difference between the best fitness of the parent population and that of the offspring population. Similar to quality gain, it can also be seen as the progress of the evolution algorithms in a single iteration [12]. The greater the gain, the faster the distance to the optimal solution decreases and the more efficient a single iteration of the optimization process will be. It is worth noting that φ_t represents the smallest fitness difference obtained by the considered EA among the first t generations instead of the counterpart at the tth generation. The purpose of this approach is to make the proposed method applicable to both nonelitist EAs and elitist EAs. The following section will show relevant cases.

Definition 5.3: Suppose that $\{\varphi_t\}_0^\infty$ is a stochastic process, where for any $t \geq 0$,

$\varphi_t \geq 0$ holds. Provided the target precision $\varepsilon > 0$, the FHT of EAs is defined by

$$T_\varepsilon = \min\{t|\varphi_t \leq \varepsilon\}$$

Especially,

$$T_0 = \min\{t|\varphi_t = 0\}$$

Furthermore, the EFHT of EAs is denoted by $E(T_\varepsilon)$.

The EFHT represents the expectation of the minimum number of iterations required for the EAs to obtain the optimal solution.

5.2.3 Problem analysis

The average gain model is a theoretical tool for runtime analysis of EAs in the continuous solution space [12,39]. However, the average gain model proposed in [39] is centered around the specific case of $(1 + 1)$ EA in a continuous domain and is not applicable for EAs widely used in practice. Zhang et al. [12] separated the average gain model from specific algorithms and objective functions to make the model more rigorous and general. However, Zhang et al. [12] analyzed the running time of EAs from a relatively abstract point of view, the average gain model will be revisited below to adapt it to the discussion on the running time of EAs in the continuous solution space. The basic framework of EAs for continuous optimization is presented in the form of pseudocode in Algorithm 1 [40]. One lemma that has been proved in [12] is presented below and it will be used to support the proof of Theorem 5.1.

Lemma 5.1: Let $\{\eta_t\}_0^\infty$ be a stochastic process, where for any $t \geq 0$, $\eta_t \geq 0$ holds. Let $T_0^\eta = \min\{t|\eta_t = 0\}$, Assuming $E(T_0^\eta) < +\infty$, if there

exists $\alpha > 0, \alpha \in \mathbb{R}$,   for   any $t \geq 0$, $E(\eta_t - \eta_{t+1}|H_t) \geq \alpha$,   then
$E(T_0^\eta|\eta_0) \geq \frac{\eta_0}{\alpha}$.

To utilize the mathematical expression designed for surface fitting of the proposed method, the conditions of [12] are slightly modified here. The domain of function $h(x)$, which was previously a closed interval, is altered to a left-open interval. As a consequence, Theorem 5.1 is revisited.

Theorem 5.1: Let $\{\varphi_t\}_0^\infty$ be a stochastic process, where for any $t \geq 0$, $\varphi_t \geq 0$ holds. Let $h:(0, \varphi_0] \to \mathbb{R}^+$ be a monotonically increasing continuous function. If $\varphi_0 > \varepsilon > 0$, when $E(\varphi_t - \varphi_{t+1}|H_t) \geq h(\varphi_t)$, then

$$E(T_\varepsilon|\varphi_0) \leq 1 + \int_\varepsilon^{\varphi_0} \frac{1}{h(x)} dx$$

Theorem 5.1 will support the use of average gain to analyze the upper bound on EFHT of EAs. $\{\varphi_t\}_0^\infty$ can be regarded as a series of historical minimum fitness differences that gradually decrease in the optimization process. Statistical experiments are used to obtain the estimated value of $E(\varphi_t - \varphi_{t+1}|H_t)$. Afterward, $h(\varphi_t)$ is obtained through surface fitting techniques. If $h(\varphi_t)$ meets the condition of Theorem 5.1, it can be utilized to derive the upper bound on EFHT $E(T_\varepsilon|\varphi_0)$. After transformation and scaling, the upper bound of running time is obtained. The theoretical basis and the implementation steps of the proposed method will be described in detail in the next section.

5.2.4 Estimation method

The overall procedure of estimating the running time of EAs for continuous optimization is provided in Fig. 5.6. The steps of the mathematical analysis approach based on the average gain model are presented in Fig. 5.6 with solid lines. The proposed method replaces the first two steps performed by mathematical analysis with two new steps assisted by experiments. The related steps are presented with dotted lines.

The research objective that includes the considered EAs and fitness functions needs to be identified before describing the proposed method. The first step of the estimation method is to collect average gain and corresponding fitness difference through statistical experiments. In the statistical experiment, the empirical distribution function derived from numerical data is utilized to simulate the probability distribution function

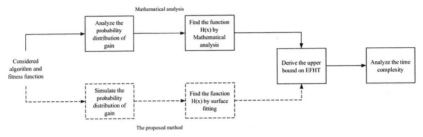

Figure 5.6 Flowchart of the proposed method.

of gain. The average gain is further calculated. Supported by surface fitting techniques, the data yielded by the statistical experiment are transformed into a function $h(\varphi_t)$ that meets the condition of Theorem 5.1.

Theorem 5.1 will be used to calculate the upper bound on EFHT of the considered EAs based on surface fitting results. It is noteworthy that these two steps are not different from the commonly used mathematical analysis methods. The difference between the proposed method and previous analysis techniques is that the proposed method replaces tedious and difficult calculations with statistical experiments and surface fitting, which significantly reduces the difficulty of analysis.

The experimental part of the proposed method will be described in detail. Fig. 5.7 shows the main steps of the statistical experiment.

First, a set of values for the parameter under consideration are selected. For example, the set {5, 10, 15, 20, 25, 30} can be chosen as the set of values for the problem size n. After the initial parameter setting, other parameters remain constant throughout the experiment.

Next, samples of gain are taken independently during the optimization process. The data representing the state of the current population are recorded, including both the minimum fitness difference ever found before generating the offspring and the counterpart after the offspring are generated. The difference between the two values is calculated to obtain the gain. Different from only generating one offspring population during one generation in Algorithm 5.1, a number of offspring populations are generated independently. Thus, a certain amount of gains are collected. The mean of these gains is calculated to represent average gains. Without loss of randomness, one of the offspring populations repeatedly generated in the iteration is selected randomly as the parent population for the next iteration. This procedure loops until the algorithm obtains the global optimal solution or the maximum number of FEs is reached.

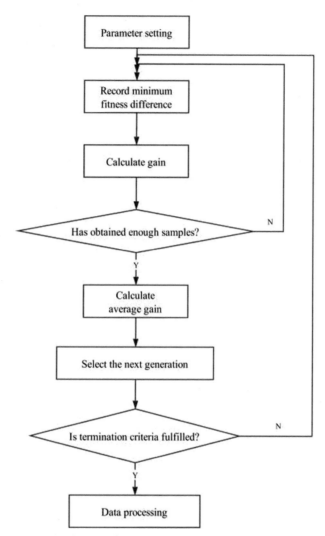

Figure 5.7 Experimental steps of the estimation method.

Notably, the samples of gain are obtained at intervals rather than during every generation. The reason is that sampling at intervals can reduce the computational cost. Moreover, taking samples during every generation would yield a large amount of sampling points. Since one sampling point corresponds to one constraint in the fitting problem, too many sampling points would render surface fitting very difficult.

To reduce the computational cost, it is necessary to select an appropriate sampling interval. Since the number of iterations required to reach the optimum varies in different scenarios, it is impossible for a common sampling interval to exist. Therefore, a pool of sampling intervals are adopted in the proposed method, for example, {50, 100, 200, 400, 800}. For each sampling interval, there will be a corresponding result and fitting error, respectively. The fitting error is divided by the number of selected data points to calculate the average fitting error. Finally, the sampling interval with the smallest average fitting error is chosen as the optimal sampling interval.

There are two cases in which some average gains of the collected sample points would be equal to 0. One of the cases is that if the sample point is not the one with the smallest fitness difference, it will be the outlier that needs to be removed because an average gain of 0 means that the algorithm cannot obtain a smaller fitness. The other case is that if the sample point is the one with the smallest fitness difference, it indicates that the algorithm under consideration falls into the local optimum. Under such a situation, there would be no need for surface fitting and further derivation. The reason lies in that if the algorithm gets trapped in the local optimum, it cannot obtain the global optimum. Hence, the average FHT will be positive infinity and there is no corresponding upper bound.

The original steps pertaining to these EAs are not modified, while several new steps for sampling are added to the algorithm. If these new steps are removed, the algorithm will still work as the original EAs.

5.2.5 Experimental analysis: simulation of gain probability density distribution function

The precise estimation of $E(\varphi_t - \varphi_{t+1}|H_t)$ of Theorem 5.1 is the vital part of the proposed method, where $\{\varphi_t\}_0^\infty$ represents a discrete non-negative stochastic process. Thus, based on Glivenko–Cantelli theorem, a statistical method is developed to estimate $E(\varphi_t - \varphi_{t+1}|H_t)$. According to Glivenko-Cantelli theorem, as the number of samples increases, the empirical distribution function will converge to its real distribution function [41]. Glivenko–Cantelli theorem is introduced below.

Let K be the number of samples and let $X_1, X_2, ..., X_K$ be the generated samples where $X_1 \le X_2 \le ... \le X_K$. Suppose that $\varphi_t - \varphi_{t+1}|H_t \sim F(x)$, where $x = \varphi_t - \varphi_{t+1}$, and that the empirical distribution function $F_K(x)$ is simulated by the statistical experiment on the basis of H_t, $F(x)$ can be estimated by $F_K(x)$ when K is sufficiently large. The formula of $F_K(x)$ is presented as follows:

$$
F_K(x) = \begin{cases}
0, & x < X_1 \\
\dfrac{i}{K}, & X_i \le x \le X_{i+1}, \ i = 1, ..., K-1 \\
1, & x \ge X_K
\end{cases}
$$

Glivenko$-$Cantelli theorem indicates that if K is sufficiently large, $E(\varphi_t - \varphi_{t+1}|H_t) \approx E(X_1, X_2, ..., X_K)$. The expectation of $\varphi_t - \varphi_{t+1}$ is approximately equal to the mean of samples $X_1, X_2, ..., X_K$. Therefore, an appropriate selection of sample size for gain in the statistical experiments will allow the mean of gains to be used to estimate the average gain.

5.2.6 Experimental analysis: fitting of the average gain surface

The purpose, restriction, mathematical expression, and methods of surface fitting will be stated in this section. There is a wide range of mathematical methods for curve and surface fitting that have been applied in many aspects of scientific research and engineering applications, such as surface chemistry [42], genome analysis [43], nuclear force field [44], etc. When aiming at introducing experimental methods into analysis models, we need to deal with the question of how to transform the experimental data into a mathematical expression for further derivation. Surface fitting methods are exactly suited for this task because they are designed to reconstruct the continuum from scattered data points by applying mathematical tools [45].

After collecting experimental data, finding a suitable function $h(\varphi_t)$ is an important step in applying Theorem 5.1 to analyze the running time of EAs. Surface fitting techniques can be used to find $h(\varphi_t)$ that meets the condition of Theorem 5.1 by fitting the surface of average gain with respect to the fitness difference and the problem size n.

It is worth noting that surface fitting techniques rather than curve fitting techniques are adopted in this part because the function obtained by curve fitting can only reflect the relationship between the average gain and the fitness difference. If curve fitting techniques are adopted, the running time analyzed by Theorem 5.1 will be only related to the termination condition and the minimum fitness difference of the initial population. However, most researchers are interested in the relationship between running time and other parameters. Therefore, it is necessary to obtain the function $f(\varphi_t, n)$ of the average gain about the fitness difference and the problem size. If the problem size takes any specific value n_i, $f(\varphi_t, n_i)$ can be regarded as $h(x)$ which satisfies the requirements of Theorem 5.1. The problem size n can be viewed as a fixed parameter when Theorem 5.1 is applied. As a consequence, the derived results can reveal the relationship between running time and problem size n.

Unlike common unconstrained conditions encountered in surface fitting, the proposed method must address surface fitting problems with constraints. In unconstrained situations, the data points are usually evenly distributed on both sides of the surface. However, since the condition of Theorem 5.1 requires $E(\varphi_t - \varphi_{t+1}|H_t) \geq h(\varphi_t)$, the value of $h(\varphi_t)$ must be smaller than the corresponding average gain pertaining to the same fitness difference and problem size. In other words, the entire fitting surface must lie below the sample points.

This section mainly focuses on the method for estimating the running time of continuous EAs. Hence, the least squares method [45] was adopted as a preliminary attempt rather than more complex methods, like moving least squares methods [46,47]. To utilize Theorem 5.1 to analyze running time, it is necessary to obtain a function that can reveal the relationship among average gain, fitness difference, and problem dimension. Although polynomial functions are widely used in the field of surface fitting [46,47], as the exponents of all items in a polynomial are positive integers, polynomial functions cannot reflect a negative correlation between the dependent variable and the independent variables directly. Considering the property of the average gain observed in the experiments, a special mathematical expression is designed for the proposed method.

The function $f(\varphi_t, n_i)$ demonstrates the mapping relationship between the average gain and the fitness difference. An intuitive idea is that the average gain is positively related to the fitness difference and negatively related to the problem size in most cases. Observing the average gain of $(1, \lambda)$ ES on the Sphere problem, it can be demonstrated that there indeed exist some situations that confirm this assumption. Based on the above conjectures and observations, the mathematical expression designed for surface fitting is as follows:

$$f(v, n) = \frac{a \times v^b}{c \times n^d}, a, c, d > 0, b \geq 1$$

where v is the smallest fitness difference and n is the problem size. Although the mathematical expression presented above is rather simple, as a preliminary attempt, it will be used for the estimation experiments presented below. Noting that for any $n_i \geq 0$, if $v = 0$, then $h(0) = f(0, n_i) = 0$. This does not satisfy the precondition of [12], which requires $h(\varphi_t) > 0$ in $[0, \varphi_0]$. That is the reason for modifying the original theorem in [12].

The parameter values in the expression need to be determined, namely a, b, c, and d in $f(v, n)$ since the mathematical expression for surface fitting is fixed. The problem of determining the parameter values is modeled as follows: given data points (n_i, v_i, z_i), $i = 1, 2, ..., m$, let $D(f)$ denote the error function, we solve the following problem:

$$\min_{f} D(f)$$

$$s.t. f(v_i, n_i) \leq z_i, i = 1, 2, ..., m$$

where n is the problem size, v is the fitness difference, and z is the average gain. The error function $D(f)$ is used to measure the deviation of the fitting results from the actual values. To reduce the distance between the fitting surface and the data points, it is necessary to minimize the error function

$$D(f) = \sum_{i=1}^{m} (\lg(z_i) - \lg(f(v_i, n_i)))^2$$

In the estimation methods, the commonly used least-squares method is slightly modified, whereby the modified error function is presented above and is defined as the sum of squared differences between the logarithmic value of average gain of each data point and that of the fitting point with the same fitness difference. The reason for this change is that the average gains of the sample points span from 10^{-10} to 10^{10} in some scenarios. If the differences among average gains are directly adopted in the original least-squares method, the effect of different points on the fitting results will not be balanced. After taking the logarithm of the average gain, the error function may take into account more equally the sample points whose fitness differences vary considerably in magnitude.

5.3 Summary

In this section, we proposed a method applying theoretical analysis results to analyze the running time of EAs for continuous optimization, including their advanced variants in practice. Statistical methods were introduced into the average gain model and sample gains to simulate the distribution function. Next, surface fitting techniques were utilized to transform samples obtained by statistical methods into mathematical expressions. The EFHT was further estimated by mathematical analysis. The proposed method does not rely on any conditions or simplifications of the considered EAs or optimization problems.

The experimental results confirm that the upper bound on EFHT of ES can be estimated correctly and effectively by the proposed method. The existing theoretical results were used to verify the estimated upper bound on EFHT of $(1, \lambda)$ ES on the Sphere problem. Moreover, the running times of ES and CMA-ES on real test functions were estimated, revealing that the obtained results are highly consistent with the numerical data. Furthermore, experiments are conducted to estimate the running time of the standard CMA-ES and an improved variant on benchmark functions.

The average gain model was taken as an example of theoretical approaches to demonstrate the utilization of statistical methods to estimate the running time of EAs for continuous optimization. Apart from the average gain model, other existing analysis tools can also be used to discuss various kinds of advanced EAs. For example, in the future, it would be beneficial to analyze the running time of advanced EAs for discrete optimization by introducing statistical methods into state-of-the-art analysis approaches, such as fitness level method, drift analysis, and switch analysis. Moreover, although we only discuss ES in this chapter, the proposed method can be applicable to other EAs, such as DE and PSO, if the designed mathematical expressions for surface fitting capture the properties of average gain accurately. This will also be further investigated in our future work. Based on the upper bound on running time discussed in this section, another direction for further research is to analyze the lower bound on running time of EAs for continuous optimization.

References

[1] Akimoto Y, Auger A, Hansen N. Quality gain analysis of the weighted recombination evolution strategy on general convex quadratic functions. Theoretical Computer Science 2020;832:42−67.
[2] Wang Y, Cai Z, Zhang Q. Differential evolution with composite trial vector generation strategies and control parameters. IEEE Transactions on Evolutionary Computation 2011;15(1):55−66.

[3] Arabas J, Biedrzycki R. Improving evolutionary algorithms in a continuous domain by monitoring the population midpoint. IEEE Transactions on Evolutionary Computation 2017;21(5):807−12.

[4] Gong W, Zhou A, Cai Z. A multioperator search strategy based on cheap surrogate models for evolutionary optimization. IEEE Transactions on Evolutionary Computation 2015;19(5):746−58.

[5] Liu Q, Chen W, Deng JD, et al. Benchmarking stochastic algorithms for global optimization problems by visualizing confidence intervals. IEEE Transactions on Systems, Man, and Cybernetics 2017;47(9):2924−37.

[6] Bartz-Beielstein T, Chiarandini M, Paquete L, et al. Experimental methods for the analysis of optimization algorithms. 1st Berlin, Heidelberg: Springer-Verlag; 2010.

[7] Jägersküpper J, Preuss M. Aiming for a theoretically tractable CSA variant by means of empirical investigations. Proceedings of the 10th annual conference on genetic and evolutionary computation. New York, NY: ACM; 2008. p. GECCO'08.

[8] Jägersküpper J, Preuss M. Empirical investigation of simplified step-size control in metaheuristics with a view to theory. In: McGeoch CC, editor. Experimental algorithms. Berlin, Heidelberg: Springer Berlin Heidelberg; 2008. p. 263−74.

[9] Jägersküpper J. Rigorous runtime analysis of the (1 + 1) ES: 1/5-rule and ellipsoidal fitness landscapes. In: International Workshop on Foundations of Genetic Algorithms, 2005. 260−281.

[10] Bäck T, Hammel U, Schwefel HP. Evolutionary computation: comments on the history and current state. IEEE Transactions on Evolutionary Computation 1997;1(1):3−17.

[11] Jiang W., Qian C., Tang K. Improved running time analysis of the (1 + 1)-ES on the sphere function. In: International Conference on Intelligent Computing, 2018. 729−739.

[12] Yushan Z, Han H, Zhifeng H, et al. First hitting time analysis of continuous evolutionary algorithms based on average gain. Cluster Computing 2016;19(3):1323−32.

[13] Jägersküpper J. Probabilistic runtime analysis of (1 + < over>, λ), ES using isotropic mutations. In: Proceedings of the 8th annual conference on Genetic and evolutionary computation, 2006. 461−468.

[14] Hansen N., Ostermeier A. Adapting arbitrary normal mutation distributions in evolution strategies: The covariance matrix adaptation. In: Evolutionary Computation, Proceedings of IEEE International Conference, 1996. 312−317.

[15] Poland J., Zell A. Main vector adaptation: A CMA variant with linear time and space complexity. In: Proceedings of the 3rd Annual Conference on Genetic and Evolutionary Computation, 2001. 1050−1055.

[16] Hansen N. The CMA evolution strategy: a tutorial. arXiv preprint arXiv 1604;00772:2016.

[17] Krause O., Arbonès D.R., Igel C. CMA-ES with optimal covariance update and storage complexity. In: Advances in Neural Information Processing Systems, 2016. 370−378.

[18] Engelbrecht A.P. Fitness function evaluations: A fair stopping condition?. In: Swarm Intelligence (SIS), 2014 IEEE Symposium on, 2014. 1−8.

[19] He J, Yao X. Average drift analysis and population scalability. IEEE Transactions on Evolutionary Computation 2017;21(3):426−39.

[20] Yu Y, Zhou ZH. A new approach to estimating the expected first hitting time of evolutionary algorithms. Artificial Intelligence 2008;172(15):1809.

[21] Chen T, Tang K, Chen G, et al. Analysis of computational time of simple estimation of distribution algorithms. IEEE Transactions on Evolutionary Computation 2010;14(1):1−22.

[22] Sudholt D. A new method for lower bounds on the running time of evolutionary algorithms. IEEE Transactions on Evolutionary Computation 2013;17(3):418−35.

[23] Wu Z, Kolonko M, Möhring RH. Stochastic runtime analysis of the cross-entropy algorithm. IEEE Transactions on Evolutionary Computation 2017;21(4):616−28.

[24] Jägersküpper J. How the (1 + 1) ES using isotropic mutations minimizes positive definite quadratic forms. Theoretical Computer Science 2006;361(1):38−56.

[25] Jägersküpper J. Algorithmic analysis of a basic evolutionary algorithm for continuous optimization. Theoretical Computer Science 2007;379(3):329−47.

[26] Beyer H, Finck S. Performance of the$(\mu/\mu I, \lambda)$-σSA-ES on a class of PDQFs. IEEE Transactions on Evolutionary Computation 2010;14(3):400−18.

[27] Beyer H, Melkozerov A. The dynamics of self-adaptive multirecombinant evolution strategies on the general ellipsoid model. IEEE Transactions on Evolutionary Computation 2014;18(5):764−78.

[28] Wegener I. Theoretical aspects of evolutionary algorithms. In: International Colloquium on Automata, Languages, and Programming, 2001. 64−78.

[29] Zhou D, Luo D, Lu R, et al. The use of tail inequalities on the probable computational time of randomized search heuristics. Theoretical Computer Science 2012;436:106−17.

[30] Witt C. Fitness levels with tail bounds for the analysis of randomized search heuristics. Information Processing Letters 2014;114(1−2):38−41.

[31] He J, Yao X. Drift analysis and average time complexity of evolutionary algorithms. Artificial Intelligence 2001;127(1):57−85.

[32] Jägersküpper J. Combining Markov-chain analysis and drift analysis. Algorithmica 2011;59(3):409−24.

[33] Chen T, He J, Sun G, et al. A new approach for analyzing average time complexity of population-based evolutionary algorithms on unimodal problems. IEEE Transactions on Systems, Man, and Cybernetics, Part B (Cybernetics) 2009;39(5):1092−106.

[34] Lehre P.K., Witt C. Concentrated hitting times of randomized search heuristics with variable drift. In: International Symposium on Algorithms and Computation, 2014. 686−697.

[35] Oliveto PS, Witt C. Simplified drift analysis for proving lower bounds in evolutionary computation. Algorithmica 2011;59(3):369−86.

[36] Yu Y, Qian C, Zhou ZH. Switch analysis for running time analysis of evolutionary algorithms. IEEE Transactions on Evolutionary Computation 2015;19(6):777−92.

[37] Yu Y., Qian C. Running time analysis: Convergence-based analysis reduces to switch analysis. In: Evolutionary Computation (CEC), 2015 IEEE Congress on, 2015. 2603−2610.

[38] Akimoto Y, Auger A, Glasmachers T. Drift theory in continuous search spaces: Expected hitting time of the (1 + 1)-ES with 1/5 Success Rule[A]. Proceedings of the genetic and evolutionary computation conference. New York, NY: ACM; 2018. p. GECCO'18. Available from: http://doi.acm.org/10.1145/3205455.3205606.

[39] Huang H, Xu W, Zhang Y, et al. Runtime analysis for continuous (1 + 1) evolutionary algorithm based on average gain model. SCIENTIA SINICA Informationis 2014;44(6):811−24.

[40] Bäck T. Evolutionary computation: toward a new philosophy of machine intelligence. Complexity 1997;2(4):28−30.

[41] Tucker HG. A generalization of the Glivenko - Cantelli theorem. The Annals of Mathematical Statistics 1959;30(3):828−30.

[42] Stiopkin IV, Weeraman C, Pieniazek PA, et al. Hydrogen bonding at the water surface revealed by isotopic dilution spectroscopy. Nature 2011;474(7350):192.

[43] Wright SI, Bi IV, Schroeder SG, et al. The effects of artificial selection on the maize genome. Science 2005;308(5726):1310−14.

[44] Xu X, Yao W, Sun B, et al. Optically controlled locking of the nuclear field via coherent dark-state spectroscopy. Nature 2009;459(7250):1105.

[45] Lancaster P, Salkauskas K. Curve and surface fitting: an introduction[M].[S.l.]. Academic press; 1986.

[46] Lancaster P, Salkauskas K. Surfaces generated by moving least squares methods. Mathematics of Computation 1981;37(155):141−58.

[47] Fleishman S, Cohen-Or D, Silva CT. Robust moving least-squares fitting with sharp features. ACM Transactions on Graphics (TOG) 2005;24(3):544−52.

Index

Note: Page numbers followed by "*f*," "*t*," and "*b*" refer to figures, tables, and boxes, respectively.